Knowledge-Driven Entrepreneurship

INNOVATION, TECHNOLOGY, AND KNOWLEDGE MANAGEMENT

Series Editor:

Elias G. Carayannis
George Washington University
Washington DC, USA

For other titles published in this series, go to
www.springer.com/series/8124

Thomas Andersson · Martin G. Curley
Piero Formica

Knowledge-Driven Entrepreneurship

The Key to Social and Economic Transformation

Springer

Thomas Andersson
Jönköping Int. Business School
Jönköping University
Jönköping, Sweden
thomas.andersson@hj.se

Martin G. Curley
Intel Corporation and
National University of Ireland
Maynooth, Ireland
martin.g.curley@intel.com

Piero Formica
Jönköping University
International Entrepreneurship Academy
Bologna, Italy
piero.formica@hj.se

ISBN 978-1-4419-1187-2 e-ISBN 978-1-4419-1188-9
DOI 10.1007/978-1-4419-1188-9
Springer New York Dordrecht Heidelberg London

Library of Congress Control Number: 2009938431

Printed on acid-free paper

Springer is part of Springer Science+Business Media (www.springer.com)

Series Foreword

The Springer book series *Innovation, Technology, and Knowledge Management* was launched in March 2008 as a forum and intellectual, scholarly "podium" for global/local, transdisciplinary, transsectoral, public–private, and leading/"bleeding" -edge ideas, theories, and perspectives on these topics.

The book series is accompanied by the Springer *Journal of the Knowledge Economy*, which was launched in 2009 with the same editorial leadership.

The series showcases provocative views that diverge from the current "conventional wisdom," that are properly grounded in theory and practice, and that consider the concepts of ***robust competitiveness***,[1] ***sustainable entrepreneurship***,[2] and ***democratic capitalism***,[3] central to its philosophy and objectives. More specifically, the aim of this series is to highlight emerging research and practice at the dynamic intersection of these fields, where individuals, organizations, industries, regions, and nations are harnessing creativity and invention to achieve and sustain growth.

Books that are part of the series explore the impact of innovation at the "macro" (economies, markets), "meso" (industries, firms), and "micro" levels. (teams, individuals), drawing from such related disciplines as finance, organizational psychology, research and development, science policy, information systems, and

[1] We define *sustainable entrepreneurship* as the creation of viable, profitable, and scalable firms. Such firms engender the formation of self-replicating and mutually enhancing innovation networks and knowledge clusters (innovation ecosystems), leading toward robust competitiveness (E.G. Carayannis, *International Journal of Innovation and Regional Development*, 1(3), 235–254, 2009).

[2] We understand *robust competitiveness* to be a state of economic being and becoming that avails systematic and defensible "unfair advantages" to the entities that are part of the economy. Such competitiveness is built on mutually complementary and reinforcing low-, medium- and high-technology and public and private sector entities (government agencies, private firms, universities, and nongovernmental organizations) (E.G. Carayannis, *International Journal of Innovation and Regional Development*, 1(3), 235–254, 2009).

[3] The concepts of *robust competitiveness* and *sustainable entrepreneurship* are pillars of a regime that we call "*democratic capitalism*" (as opposed to "popular or casino capitalism"), in which real opportunities for education and economic prosperity are available to all, especially – but not only – younger people. These are the direct derivative of a collection of top–down policies as well as bottom–up initiatives (including strong research and development policies and funding, but going beyond these to include the development of innovation networks and knowledge clusters across regions and sectors) (E.G. Carayannis and A. Kaloudis, *Japan Economic Currents*, p. 6–10 January 2009).

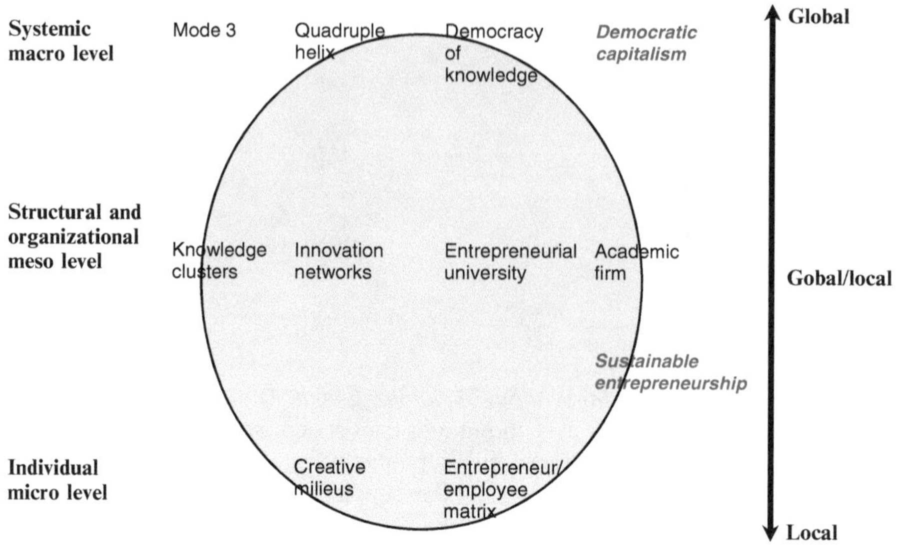

strategy, with the underlying theme that for innovation to be useful it must involve the sharing and application of knowledge.

Some of the key anchoring concepts of the series are outlined in the figure below and the definitions that follow (all definitions are from E.G. Carayannis and D.F.J. Campbell, *International Journal of Technology Management*, 46, 3–4, 2009).

Conceptual profile of the series *Innovation, Technology, and Knowledge Management*

- The "Mode 3" Systems Approach for Knowledge Creation, Diffusion, and Use: "Mode 3" is a multilateral, multinodal, multimodal, and multilevel systems approach to the conceptualization, design, and management of real and virtual, "knowledge-stock" and "knowledge-flow," modalities that catalyze, accelerate, and support the creation, diffusion, sharing, absorption, and use of cospecialized knowledge assets. "Mode 3" is based on a system-theoretic perspective of socioeconomic, political, technological, and cultural trends and conditions that shape the coevolution of knowledge with the "knowledge-based and knowledge-driven, global/local economy and society."
- Quadruple Helix: Quadruple helix, in this context, means to add to the triple helix of government, university, and industry a "fourth helix" that we identify as the "media-based and culture-based public." This fourth helix associates with "media," "creative industries," "culture," "values," "life styles," "art," and perhaps also the notion of the "creative class."
- Innovation Networks: Innovation networks are real and virtual infrastructures and infratechnologies that serve to nurture creativity, trigger invention, and

catalyze innovation in a public and/or private domain context (for instance, government–university–industry public–private research and technology development coopetitive partnerships).

- Knowledge Clusters: Knowledge clusters are agglomerations of cospecialized, mutually complementary, and reinforcing knowledge assets in the form of "knowledge stocks" and "knowledge flows" that exhibit self-organizing, learning-driven, dynamically adaptive competences and trends in the context of an open systems perspective.
- Twenty-First Century Innovation Ecosystem: A twenty-first century innovation ecosystem is a multilevel, multimodal, multinodal, and multiagent system of systems. The constituent systems consist of innovation metanetworks (networks of innovation networks and knowledge clusters) and knowledge meta-clusters (clusters of innovation networks and knowledge clusters) as building blocks and organized in a self-referential or chaotic fractal knowledge and innovation architecture (Carayannis 2001), which in turn constitute agglomerations of human, social, intellectual, and financial capital stocks and flows as well as cultural and technological artifacts and modalities, continually coevolving, cospecializing, and cooperating. These innovation networks and knowledge clusters also form, reform, and dissolve within diverse institutional, political, technological, and socioeconomic domains, including government, university, industry, and nongovernmental organizations and involving information and communication technologies, biotechnologies, advanced materials, nanotechnologies, and next- Generation energy technologies.

Who is this book series published for? The book series addresses a diversity of audiences in different settings:

1. *Academic communities:* Academic communities worldwide represent a core group of readers. This follows from the theoretical/conceptual interest of the book series to influence academic discourses in the fields of knowledge, also carried by the claim of a certain saturation of academia with the current concepts and the postulate of a window of opportunity for new or at least additional concepts. Thus, it represents a key challenge for the series to exercise a certain impact on discourses in academia. In principle, all academic communities that are interested in knowledge (knowledge and innovation) could be tackled by the book series. The interdisciplinary (transdisciplinary) nature of the book series underscores that the scope of the book series is not limited a priori to a specific basket of disciplines. From a radical viewpoint, one could create the hypothesis that there is no discipline where knowledge is of no importance.
2. *Decision makers – private/academic entrepreneurs and public (governmental, subgovernmental) actors:* Two different groups of decision makers are being addressed simultaneously: (1) private entrepreneurs (firms, commercial firms, academic firms) and academic entrepreneurs (universities), interested in optimizing knowledge management and in developing heterogeneously composed knowledge-based research networks; and (2) public (governmental, subgovernmental) actors that are interested in optimizing and further developing their

policies and policy strategies that target knowledge and innovation. One purpose of public *knowledge and innovation policy* is to enhance the performance and competitiveness of advanced economies.

3. *Decision makers in general:* Decision makers are systematically being supplied with crucial information, for how to optimize knowledge-referring and knowledge-enhancing decision-making. The nature of this "crucial information" is conceptual as well as empirical (case-study-based). Empirical information highlights practical examples and points toward practical solutions (perhaps remedies), conceptual information offers the advantage of further-driving and further-carrying tools of understanding. Different groups of addressed decision makers could be decision makers in private firms and multinational corporations, responsible for the knowledge portfolio of companies; knowledge and knowledge management consultants; globalization experts, focusing on the internationalization of research and development, science and technology, and innovation; experts in university/business research networks; and political scientists, economists, and business professionals.
4. *Interested global readership*: Finally, the Springer book series addresses a whole global readership, composed of members who are generally interested in knowledge and innovation. The global readership could partially coincide with the communities as described above ("academic communities," "decision makers"), but could also refer to other constituencies and groups.

Elias G. Carayannis
Series Editor

Preface

We are living in a time of extraordinary change. At this juncture, there are opportunities for radical improvement and solutions to our problems, but we are also faced with dramatic challenges and threats to our very existence. Both hopes and problems often center on the role of technology and what technology brings. Today, however, our future no longer depends on what technology can or cannot do for us. Eventually, the advancing tools of new technology will do basically what we require them to do. Today, and tomorrow, it is all about people, and what we set out to achieve – as individuals and collectively.

The key is knowledge and how we put knowledge into action. And because knowledge can be invested in so many activities, its role has become greater than ever before. Knowledge cannot be taken for granted, nor can its use. Knowledge may be disruptive, it may deteriorate, and it may be destroyed or just quietly pass into oblivion. Knowledge and its sensible management are here for us to cherish.

No one-size-fits-all approach exists. And knowledge cannot be measured on a single scale. There are different kinds and ways of using it, which in part reflect people's varying experience. Unique and novel constellations can come about as people with different backgrounds and experiences come together and complement each other.

In producing this book, we have had the advantage and privilege of enjoying the collaboration with colleagues who are able to draw on a diverse range of backgrounds but who have shared a common cause. Our backgrounds span academia, policymaking, and business. We were born and have worked in different countries and types of organizations. We have different experiences of working with decision-makers, ranging from international to national, regional and local level. Our hope is that the book and its diverse perspectives will serve as a source of inspiration for people from different backgrounds in different parts of the world but who share a common concern that we need to move toward models that are more conducive to development and the use of knowledge.

Jönköping, Sweden
Dublin, Ireland
Bologna, Italy

Thomas Andersson
Martin Curley
Piero Formica

Contents

Introduction

Entrepreneurship is the glue that holds the knowledge triangle made up of education, research and innovation, which are key drivers of a knowledge-based society.

Martin Schuurmans
Chairman of the European Institute
of Innovation and Technology

The Knowledge Age: A Golden Future in Sight?

We have entered a new age heralded by the "enlightenment" of the "knowledge in action" revolution. Much has been written and spoken about the associated accumulation of production factors (i.e., capital, labor, and raw materials) that could help explain long-term growth potential.

The newly coined interest in knowledge as a production factor is, however, generally approached from too restrictive a viewpoint. Knowledge generation is not synonymous with scientific advances. Knowledge in action is not necessarily about diffusing scientific discovery or creating technology-based high-growth companies. Neither is knowledge a "new" factor of production.

Peter Drucker's and William Baumol's focus on knowledge and entrepreneurship, respectively, has rather thrown the spotlight on the importance of combining inputs in profitable ways. Knowledge creation and knowledge use are fuelled by the combined influence of human creativity and entrepreneurial energy channeled into innovation (Baumol 2002) in ways that enable successful, unforeseen combinations of different production factors. In turn, the effectiveness of the instrumental role played by knowledge creation is attributable to the culture of the institutions involved. An institutional environment that encourages creativity and experimentation is the ultimate determinant of economic growth and renewal.

The flow of knowledge, not technology per se, is the hallmark of technology transfer. To be effective, technology transfer therefore needs trustful and outward-looking knowledge brokers. The fact is that the knowledge revolution, caught up in its own enormous success, has in effect contributed to the rise of an economy which, in its present format, has proven unsustainable. Major problems exist in terms of reliability and accountability in business, while governments and multilateral institutions are failing to provide satisfactory cross-border policy frameworks in a range of areas. Finally, it is important that market transactions and innovations are not merely pushed by technocrats and experts, but that they are pulled by the real needs of people, and of society, to produce better responses to real issues.

Knowing what a technology means in all its facets and how it can be transferred from its source to the user is the broker's main role. This implies that the broker should be a "cross-pollinator" who handles a chain of interactions between different professional communities of knowledge practice (engineers, patent and business lawyers, business economists, distribution agents, etc.) in various cultural contexts and countries. Care also must be taken of the interstices between them.

In a context of strong trusting relationships that are cross-functional, cross-sectional, and across borders, a threat for today's technology transfer intermediaries is that of foreordaining a conduct very similar to the behavior of professional corporations embedded in nation-states. Knowledge innovation communities, which embody the principles of the free-market economy and whose application-minded agents can relate themselves more effectively to the knowledge economy than the corporatist groups, are on sustained trends of growth improvement.

The needs of knowledge innovation communities and those of corporatist associations often do not match. The value proposition of knowledge policy is that it helps shape a new market context in which communities of free knowledge-innovation agents take a powerful step toward building a knowledge-intensive economy. Knowledge-relevant economic policy is that which empowers knowledge innovation agents who break away from the prevailing trend of corporatist routines seen from the medieval guilds up to today's professional corporations. These agents are interested in a type of policy relevant to the stimuli to change in accordance with new ideas that knowledge provides. They go with the flow of knowledge.

Economic policy that endorses corporatist pressures and disregards market signals cannot accommodate the higher potential of the knowledge economy. Professional corporations engrossed in their bureaucratic practice and attitude are not apt to seize the broader field of vision that the knowledge economy presents or to spot the interdependence of the various branches of knowledge.

Again, modern governance-building institutions have failed to put in place the prerequisites for fulfilling the potential of the knowledge era. Consequences in terms of failure are visible in a number of respects:

- Lack of an orderly framework for protecting intellectual property rights and thus for accurately measuring and disclosing the value of intangible assets, to the detriment particularly of individual innovators and SMEs with limited bargaining power.
- Within organizations, lack of leadership by management to take charge of information management. This results in weaker rights, obligations, and strategic leadership and lack of accountability, which is visible in multiple spheres – business, politics, universities.

- A "shallowness" in many activities, a thinning of professional skills and an expansion of meaningless or even destructive services, as observed in areas ranging from legal services to manual work.
- Lack of verifiability and accountability in data governance, feeding cybercrime and fraud, and also contributing to weakening oversight whose consequences are all too visible in the performance of the financial services industry over the last decade.
- When there is lack of orderly conditions there is also failure to observe and take action to preserve and manage resources that generate common goods for the long term, but whose protection requires investment in the short term. Many societal and environmental assets fall into this category. They include inherited cultural assets that will become more valuable when preserved for future generations (for instance, the upbringing of our young, whose full voice will be heard only later on) and critical environmental assets such as the air we breathe and the water we drink.

The implications for public policy are clear. Policymakers have a distinct role to play in creating and ensuring the presence of sound institutions that can support effective markets. Corporatist policy, on the other hand, is damaging to the efficient operation of knowledge markets and hence to the knowledge economy as a whole. Incumbent organizations become vested interests. They try to boost embedded costs and what they already take pride in. Genuine experimentation and renewal requires openness of product and factor markets for newcomers to make entry.

Following this view, the struggle of the open economy, which is a vehicle of knowledge innovation, against the closed economy, which debars the flow of energy, remains at the heart of the debate on knowledge-relevant economic policy.

Entrepreneurship in the Knowledge Age

Entrepreneurs of different nationalities and backgrounds are creating start-ups, which are instrumental to the development of the knowledge age. They build "alumni" communities of entrepreneurs that combine their efforts to overcome the limitations of the established national boundaries and mindsets with the aim of igniting and propelling the flow of knowledge to form international start-ups.

This growing phenomenon is rendered possible by globalization, human mobility, and by the increasing force of the so-called creative class. It is further propelled by the obvious advantages it conveys, given orderly conditions. Naturally it need to be accompanied by orderly regulatory frameworks in support of transparent markets and financial transactions, societal and environmental protection, competition law, orderly multilateral frameworks to underpin intellectual property rights, etc. Again, a lot of this is simply not in place. We are facing a crisis when it comes to the ability of governance mechanisms to deliver on many of our most important issues.

We cannot expect the answers to come from to fall like manna from heaven. Unlike in the economics textbooks, there is no altruistic government that looks to the common good of society as a whole. The policies of government are well known to reflect political processes, where vested interests exercise disproportionate influence and the timing of elections influence when popular (and unpopular) decisions are made. Scientific discovery, and new knowledge more broadly, can of course translate into public opinion and thereby influence government policy. That link is not exogenously given, but – in the knowledge economy – any government with an ambition to take and implement decisions needs a thorough communication policy, as does any private company or university for that matter.

The way in which new knowledge is diffused in media and society until it translates into generally accepted knowledge is an area of study in itself, however, subject to its own logic which in turn are changing as the information revolution changes the media and as we devote our attention to different media. The traditional monopolistic position of the public TV channel is now all but gone, and with the convergence of the Internet with cellular technology and other means of communication, we are destined to witness increasingly cutthroat competition. The view of what constitutes reliable and attractive sources is in a state of flux. The readiness of the media to highlight a particular topic is also not a given, as it will depend on what can be "sold" as news. On this basis, there is today a systematic tendency for the sudden event to take preeminence over creeping realization. There is the pressure for all the actors – business people, analysts, financial journalists, newspaper owners, TV channel owners, and so on – to share in this distorted presentation of how the world evolves.

But while the impulses for action and for knowledge generation cannot in a general sense come "from above," a fundamental feature of the knowledge era is the scope it creates for bottom–up initiative. Here, entrepreneurship is a key aspect.

Widespread entrepreneurial activity is a universal necessity. This was always the case and is so more than ever. Entrepreneurialism and technology are central to a country's economic growth and social health. A culture of entrepreneurship helps to diversify the economy and make it more entrepreneurial. Entrepreneurial economies – those with a high rate of conversion of new ideas into fast-growing enterprises – display superior macroeconomic performance.

The quality of company formation is crucial. The difference between small and entrepreneurial businesses is ultimately attributable to innovation. Policy actions have to be conducive to the creation of innovation-driven firms.

A direct link exists between the entrepreneur and the innovation process. Innovation is knowledge turned into action through creative endeavor that rests heavily on the willingness of individuals to start new firms. International entrepreneurship, which takes advantage of globalization, accelerates this process by increasing the opportunities for the successful commercialization of innovation. From the supply side, innovative goods and services can be extended to new geographical markets. From the demand side, diversity in demand (i.e., increased demand for tailor-made and individualized goods and services that are the outcome of cross-cultural influences and regional, social, and cultural differences) can be satisfied.

Attitudinal differences related to both local and individual cultural features result in divergent performances when building and sustaining cross-border and cross-cultural firms. As with international entrepreneurship, the world is our community. Intellectual exchanges and the mobility of human capital are qualifying elements of the international dimension of entrepreneurship. In particular, student mobility – a notion that is at the same time new and traditional – is the product of matchmaking by education institutions in different countries. These institutions should reflect more thoroughly on how to respond to the new meanings of entrepreneurship and to the new challenges that the start-ups incubated in their environments will face in the years to come.

Knowledge communities are central to the formation of a culture of international entrepreneurship that helps to diversify the economy and make it more entrepreneurial. In low-context communities, interpersonal collaboration across multiple boundaries – across cultures, functions, rivalries, geography – creates the right atmosphere for the rise and spread of broader and fresher insights from newcomers. In this respect, all sorts of organizations, supported by policy interventions and integration, must encourage both physical and virtual mobility across national, cultural, and ethnic boundaries.

Policymakers must take action to help fuel international entrepreneurship. A borderless entrepreneurial economy leads to progress, development, and prosperity for citizens and society as a whole. Society and its actors – government, public and private organizations, universities, and higher education institutions, companies, and political and economic development leaders – are called upon to participate in the creation of the next breed of entrepreneurs in a more active, direct, and leading capacity. All actors must be committed to developing an entrepreneurial culture, raising students' awareness of entrepreneurship as one of many alternative career paths, encouraging cross-border and intercultural entrepreneurial behavior, and boosting international start-up undertakings.

Because potential entrepreneurs face the choice of whether, and with whom, to collaborate when setting up their own business, they need to take account of the impact that an international team of founders would have on the likely success of their new business.

International teams are better suited to competition in the global marketplace because international entrepreneurship is a mechanism that makes available the right partners required by nascent entrepreneurs to establish relationships that increase the flow of ideas and provide the institutional settings conducive to the formation of multinational and multicultural teams of entrepreneurization.

This book introduces and emphasizes the relevance and importance of high-expectation entrepreneurship as a driver of economic growth. Entrepreneurial ventures differ dramatically from the process of expanding existing businesses since little is known and much has to be assumed. Using a translational innovation approach, the book borrows paradigms from experimental research and medical practice as vectors for accelerating the experimentation and validation of business models associated with high-expectation entrepreneurship. The ability to transpose, test, and iterate new business ideas and models in a laboratory environment has significant potential. Emerging software platforms, which can simulate markets and consumer reactions or help predict demand for new products, when coupled with approaches such as

Discovery Driven Planning (McGrath and McMillan 1995), can lead to rapid learning, iteration, and preliminary validation of a new business idea, risk mitigation, avoidance of potential costs, and maximization of revenue potential.

We are all on the move into the era of the knowledge economy. Knowledge was always vital to human activity, but now its role is becoming pervasive. How knowledge is developed and put to use is critical for setting the stage in terms of the progress and prosperity that countries and economies can achieve. Arguably countries and regions need new socio-economic models, one's which are sustainable and do not depend on ever increasing growth and consumption of resources. Knowledge Driven Entrepreneurship may be a key contributor to the development of such models and their effective operation thereafter, success cannot be taken for granted. Outcomes will critically depend on the extent to which societies and institutions are able to create conditions that enable innovation and entrepreneurship.

References

Baumol WJ (2002) The free-market innovation machine. Princeton University Press, Princeton, NJ
McGrath RG, McMillan I (1995) Discovery driven planning. Harvard Bus Rev 73(4):44–54

Notes

[1] We define *sustainable entrepreneurship* as the creation of viable, profitable, and scalable firms. Such firms engender the formation of self-replicating and mutually enhancing innovation networks and knowledge clusters (innovation ecosystems), leading toward robust competitiveness (E.G. Carayannis, *International Journal of Innovation and Regional Development*, 1(3), 235–254, 2009).

[2] We understand *robust competitiveness* to be a state of economic being and becoming that avails systematic and defensible "unfair advantages" to the entities that are part of the economy. Such competitiveness is built on mutually complementary and reinforcing low-, medium- and high-technology and public and private sector entities (government agencies, private firms, universities, and nongovernmental organizations) (E.G. Carayannis, *International Journal of Innovation and Regional Development*, 1(3), 235–254, 2009).

[3] The concepts of *robust competitiveness* and *sustainable entrepreneurship* are pillars of a regime that we call "*democratic capitalism*" (as opposed to "popular or casino capitalism"), in which real opportunities for education and economic prosperity are available to all, especially – but not only – younger people. These are the direct derivative of a collection of top–down policies as well as bottom–up initiatives (including strong research and development policies and funding, but going beyond these to include the development of innovation networks and knowledge clusters across regions and sectors) (E.G. Carayannis and A. Kaloudis, *Japan Economic Currents*, p. 6–10 January 2009).

Part I
Knowledge

Chapter 1
Scenario Setting

The idea that knowledge is something anyone can acquire is recent: for most of history knowledge has been rare and secret, and this esoteric heritage, with its dream of mastery and mystery, survives in the jargon with which every profession protects itself. Knowledge is still a serpent eating its own tail.

(Theodore Zeldin)

We live in the fast-moving times and discontinuous changes of the new age of knowledge, information and access (see Appendix 1), when nimbleness, speed, transparency and local sensitivity have become absolutely essential to success.

The transition from the machine-age to the knowledge-age has raised an awareness of the need to replace the old guard of atom-based firms with a new breed of counter-cultural, digital-rooted companies. They employ brains instead of hands, invest in new concepts and contents rather than in new machines, accept fast change as a constant, and behave as coopetitors who collaborate and compete with other companies at the same time to generate new market spaces, instead of fighting for existing markets.

The Industrial Age relied upon scarce resources. Land, labor and capital were the drivers of progress in the industrial economy, with knowledge essentially of the type incorporated in machines and other tangible or physical assets. The knowledge economy relies upon the asset of human knowledge. In the knowledge economy, the main source of value creation resides in the hyper-mobility of knowledge, which controls access to opportunity and advancement (see Appendix 1 for prima facie evidence of a divide between the Industrial Age and the knowledge economy).

Knowledge is a human process dealing with mental objects, and requires awareness and intuition. In today's most technologically advanced countries, knowledge is the primary factor determining the standard of living.

T. Andersson et al., *Knowledge-Driven Entrepreneurship*, Innovation, Technology, and Knowledge Management 102, DOI 10.1007/978-1-4419-1188-9_1,

Exhibit 1: Knowledge Economy

A knowledge-based economy is focused on generating and using knowledge to create value through activities such as:

- Investments in R&D
- Leveraging emerging technology directions (see Appendix 2)
- Patenting
- Development of scientists and engineers

Some of the more advanced applied thinking about the instantiation of a knowledge economy has been performed by the European Commission. Knowledge lies at the core of the European Union's Lisbon Strategy to become the "most dynamic competitive knowledge-based economy in the world." A knowledge triangle comprising research, education and innovation has been conceptualized as a core factor in European efforts to meet the ambitious Lisbon goals. Numerous programs, initiatives and support measures are carried out at the EU level in support of knowledge. The EU Seventh Framework Programme (FP7) bundles all research-related EU initiatives together under a common program playing a vital role in attempting to reach the Lisbon strategy goals of growth, competitiveness and employment. This FP7 program is one of the world's largest open innovation programs creating research priorities, infrastructure, resources and funding to enable a knowledge economy. Although at the time of writing the EU is well behind its objective of reaching the Lisbon goals, the Lisbon strategy remains one of the more ambitious and well-conceived plans to enable a knowledge economy across a broad geographic region.

A sustainable knowledge economy is achieved only when entrepreneurs create products and services that are useful and value-adding and delivered at price points that participants (consumers, enterprises, and public sector bodies) in different markets are able to afford and willing to purchase. Indeed in the knowledge economy, consumers may transition to become prosumers actively participating in the creation of knowledge and become an integral part of the knowledge chain process (ISTAG 2009) with the social production elements involved in YouTube and Wikipedia being present day mainstream examples of this.

A key opportunity for the knowledge economy is the concept of "dematerialization," where as the cost of information continues to fall it is increasingly substituted for other resources in the physical value chain such as labor and raw materials (McInerney and White 2000). Dell's use of a streamlined IT-enabled value chain allows it to build PCs to order rather than making to stock, avoiding building costly inventory which has a short half-life. Likewise the provision of automated answering systems and online self help for IT services and products replacing human operators is another example of substituting bits for atoms.

The broader use and availability of knowledge and the ability to move it more quickly will enable key societal shifts such as moving from resource intensive

(continued)

Exhibit 1 (continued)

living to a sustainable lifecycle, and moving from geographically-based communities to virtual communities (ISTAG 2009)

Much of the growth in labor productivity outside the agricultural sector is concentrated in knowledge-intensive activities, with business and ICT services currently accounting for the bulk of labor productivity growth (OECD Science, Technology and Industry – 2003, Scoreboard, http://www.oecd.org/sti/scoreboard).

Indeed, a big ongoing change in business is the attempt to move to service models. Companies that historically were producers are trying to move away from a one-time purchase to an annuity stream by providing a service. And inevitably that is enabled by IT. Rolls-Royce, which is now trying to sell "power by the hour" is a prime example. Instead of selling an aircraft engine and maybe a maintenance contract on top of that, Rolls-Royce can use automated telemetry to offer something that is completely different.

Increasing competition for the factors that generate knowledge is accompanied by rapid integration of the Internet into everyday life, and global integration of economic activities.

Knowledge is not information. The latter is an inert resource or a static activity of reading, duplicating and broadcasting news. The former is a purposeful and dynamic process of selection and interpretation of information, and of face-to-face interactions through which knowledge is continuously recreated and meanings are assigned to facts that otherwise would remain unintelligible. There is no information without rendering it explicit. Conversely, there is knowledge although not explicit (Sveiby 1997).

The producer of knowledge, unlike the manufacturer of a physical product, still keeps it, once knowledge has been surrendered in exchange for money. This raises two points.

From one point of view, knowledge goods are, in economic parlance, "nonrival" (that is, they can be used by their vendors and buyers simultaneously). "Knowledge – as knowledge experts say – is not given up in exchange for money in the same way as a cream cake. You can't eat your cake and have it, but you can sell your knowledge and keep it" (Hampdem-Turner and Trompenaars 1994). Conversely, things made up by mass or energy are "rival" in that they cannot be used by two or more persons at once (Bailey 2000).

From another point of view, knowledge is not a limited-resource market like agriculture, mining and bulk goods, where a fixed-resource constraint is put on their trading. Knowledge markets are not affected by a short supply of ideas: the potential for finding new ideas is infinite.

Exhibit 2: Distinctive Attributes of Knowledge and Information

Knowledge	Information
Mental tools that make sense of things	A message that reduces uncertainty
An evolving set of beliefs about the world	
A crucial production factor that changes old Knowledge makes mere information valuable	
Dynamic	Static
Dependent on individual	Independent of individual
Tacit	Explicit
Analogue	Digital
Must be recreated	Easy to duplicate
Face-to-face communication	Easy to broadcast

Source: Leonard (1998), Sveiby (1997)

Exhibit 3: Rival and Nonrival Goods

Paul Romer, the founding-father of the new growth theory, divides the economy into ideas or "nonrival" goods, which can be stored in a piece of string, and things or "rival" goods with mass or energy. For example, cars are rival goods; recipes, formulas and techniques used to rearrange things are nonrival goods.

Increasingly, the knowledge economy has moved in a direction in which at least some kind of knowledge is put to work (in tools, processes and products, and in knowledge itself) for innovation with a short life-cycle.

Exhibit 4: Innovation, Innovation Life Cycle and Knowledge Innovation

Innovation, the tool of the entrepreneur (Drucker 1993a), is the introduction of something new which creates value for the organization that adopts it (Curley and Baldwin 2007).

Traditional perspectives have viewed innovation as closely related to science and technology. Mastering the expanding opportunities in scientific and technological progress is indeed becoming an increasingly important source of generating high value-added innovation. On the other hand, innovation can take many forms, including commercialization of science and technology as well as the development and implementation of new ideas more generally, as in the form of organizational change or inventing new ways of doing things. Innovations that enhance attractiveness and accessibility to customers and users are often essential

(continued)

Exhibit 4 (continued)

for commercialization. Further, innovation is the key not only to economic progress, but also to identifying new solutions to pressing social issues, such as an ageing population or environmental degradation. Innovations may be categorized in different ways, including product and process innovations, although there is no clear-cut dividing line between the two.

Innovation must not be conceptualized as a one-dimensional, linear process leading from certain input factors. Innovation is the result of efforts by multiple actors, and is enhanced by their constructive interactions. No single actor generally manages all the skills that are useful, but complementary competencies are crucial, allowing a continuous flow of impulses from both the supply and the demand side to meet. Fostering conditions that are favorable to innovation may require reforms in various areas.

The outcome of interactions that are conducive to the exchange of knowledge, innovation arises from ongoing circles of exchange where information is not just accumulated or stored, but created.

Therefore, innovation is knowledge in action that translates something originating in an experiment into something newly introduced in the marketplace. It leads to changes in the way of doing things that are perceived as positive by those organizations or individuals who make use of it.

The innovation life cycle is represented by an S-shaped logistic curve consisting of the three distinct phases that – as has been illustrated, for example, by Howard and Guile 1992: 12 – refer, respectively, to "emergence (the development of the product or service, its manufacturing capabilities, and its place in the market), growth (where the product family pervades the market), and maturity (where the market is saturated and growth slows)."

As George Kozmetsky has observed, "The life cycles of economic goods and services that are digital in form and heavily dependent on knowledge are often short, due to intense product development. As new and advanced products are launched in the market, the earlier generations become obsolete. Typically, the new generation of a product embodies not only upgraded technological and marketing characteristics, but also a wider array of attributes. Defining a product by the vector of services it delivers, the dimensionality of this vector increases all the time. Products become more complex" (Voice of the Entovation 100, 2002).

Knowledge is generated anew from connections that were not there before. Knowledge Innovation®[1] refers to the creation, evolution, exchange and application of new ideas into marketable goods and services for (1) the success of an enterprise, (2) the vitality of a nation's economy; and (3) the advancement of society (Amidon 2001).

Disruption is the essence of knowledge innovation. In fact, it establishes entirely new compelling performance trajectories whereby forays into new market segments or new businesses look very attractive. Disruptive innovation sweeps away the traditional competitors whose products or services are hit by irreversible

(continued)

Exhibit 4 (continued)

obsolescence. Customers become aware of new commercial opportunities that stretch beyond their current requirements (Christensen 1997).

Knowledge Innovation is fostered by information gathered from new connections; from insights gained by journeys into other disciplines or places; from active, collegial networks and fluid, open boundaries – which give birth to knowledge innovation communities.

Industrial culture focuses mainly on the production of "things" – of static objects. Knowledge, on the other hand, is constantly in flux, like a flowing stream by which markets are reformed and redefined. At the same time, skills oriented toward the former have a tangible target, and may be based on high levels of professionalism. Skills for the latter may be more evasive.

Exhibit 5: Reformed and Redefined Markets

Reformed markets are the results of the reformulation of existing ideas. Technologies do not change the basic structure and functioning of the market: they help to squeeze out costs and facilitate interactions. They are improvements rather than a wholesale redefinition of the R&D, marketing and sales processes, supply chains, and so on.

Incumbents have built-in advantages: a trusted brand name, an established reputation, established customer relationships, financial depth and deep pockets. Despite their strengths, they suffer from the disadvantage that their resources, strategies and structures do not allow them to envisage revolutionary possibilities.

Redefined markets are created when market boundaries and norms are redefined and an entirely new market emerges. Redefined markets are the realm of inventors, independent innovators and associated entrepreneurs. They create new firms that bring innovation to the marketplace.

Refined markets endanger the status of the incumbent entrepreneurs embedded into reformed markets, for the revolutionary business opportunities envisioned by the new firms cannot be encompassed with the incumbents' range of resources, strategies or structures.

Source: Day and Fein (2003), Baumol (2002)

As Peter Drucker elucidated, "Knowledge has two incarnations: knowledge applied to existing processes, services, and products is productivity; knowledge applied to the new is innovation" (Drucker 1997).

Conventional industrial notions lead policymakers to believe that the addition of a knowledge-based industry to an existing industrial base makes a knowledge economy. This is not the case. Pieces of knowledge, purchased like objects, do not make a knowledge economy. What is missed in such perceptions is the importance:

- of managing and synthesizing knowledge to create value
- of conducting conventional business in innovative ways

In the knowledge economy the marketplace is not divided into towns and regions but into affinity groups (see Exhibit 37) that emerge from a high propensity to sociability (also known as "invisible networks of peers" –Carayannis and Allbritton 1997).

In a truly open global economy, no single country is able to dominate others in isolation: knowledge-driven economies and knowledge-based societies can materialize only through the "chemistry" of community.

The transition to such a state of social, political and economic affairs is full of challenges as well as opportunities. Basically all economies struggle to capture the potential benefits of the modern-day knowledge society, economy and polity.

The path to a new age of prosperity is full of pitfalls that can trigger socioeconomically regressive trends and patterns (from *nouveaux pauvres* to fundamentalists of all hues, including neo-Luddites – see Carayannis, GWU Lectures 1996–2005).

We are still in transition from the decline of a mass-production economy and the rise of value from knowledge. Value from efficiency and scale is no more a competitive differentiator. The major significance of the new age is value from innovation that comes from the life-blood of knowledge in action.

As more commercial activities shift toward knowledge and information, the economy has evolved toward rapidly increased volumes of global trade bits. Ever more demanding individuals and companies have been willing not just to test the waters of knowledge markets but to take the plunge. Eventually, under conditions of macroeconomic laxness and regulatory retreat, the financial markets have adopted new instruments and expanded under conditions lacking transparency and accountability. The service sector more broadly has inflated beyond limits and is, as of 2009, in for a period of sharp consolidation.

Knowledge markets are nevertheless poised to continue expanding their role as an engine for economic development. They are likely cast in the role product markets played all through the industrial era. That said, their role is not separate from that of products and services in a traditional sense: it relates to and helps leverage those of basically all other products.

Knowledge is transmitted to markets through a variety of ideas embodied in recipes, formulas and techniques, whose common denominator is their endless capacity to rearrange physical objects from low to high value configurations. For example:

- From silicon used primarily to make glass to its use as a crucial component in microchips and optical fibers.
- From wood-intensive campfires to clean, efficient natural gas for cooking food.
- From heavy earthenware pots to ultra-thin plastics and lightweight aluminium cans (Bailey 2000).

There are different types of "knows" like those listed below:

- Know-how: skills in managing practical processes, which means selling the knowledge of how information must be processed.
- Know-what: knowledge about facts – that is, what sort of information is needed.
- Know-why: explanatory science – that is, why a given type of information is needed.
- Know-where: where information can be founded to achieve a specific result.

- Know-when: by what time information is needed.
- Know-who: knowledge about socially related understandings (van der Speck and Spijkervet 1997; Reid 2001: 5).

Exhibit 6: Knowledge Markets

Knowledge markets are a conceptual market space in which bits are traded across the continents in a broad range of content that would include:

- Knowledge and information systems
- Customer knowledge and support
- Knowledge arbitrage and exchange
- Expert exchange
- E-learning exchange
- Intellectual property
- Economic and business intelligence

Knowledge markets are somewhat removed from the competitive end of the scale that measures what influence, if any, producers can exert on prices, and how great that influence may be. These markets are characterized by imperfect or monopolistic competition (Robinson 1960: 222–245); – that is, a contextual presence of elements of competition and monopoly that intermingle wide-rangingly, depending on both the heterogeneity of what the monopolistic competitors offer and the preference that customers show for one firm over another. Indeed, heterogeneity is a distinctive feature of knowledge markets. In most of these, competitors are completely differentiated once they create ideas that modify the functions performed by finite resources so as to make a broader spectrum of possible solutions available to potential buyers.

Yet, imperfection of competition upstream at the knowledge-creation and production stages fosters more competition downstream at the point of distribution. End-users can profit not only from greater choice and a wider variety of outlets, but also from significant price reductions which competing vendors can sustain through market growth.

The Internet, undersea fiber-optic cables and communications satellites are the infrastructures that enable access to knowledge markets. Additionally the emergence of disruptive mobile broadband technologies such as WiMAX enable access for a much broader range of people.

A great variety of offers (richness) and the amplitude of connectivity (reach) give the participants unlimited capacity to weave relationships and profit from their advancements.

As more commercial activities shift toward knowledge and information, the economy seems ripe for an increase in the volume of those elements that are globally traded. Ever more demanding individuals and companies will be willing not just to test the waters of knowledge markets, but to take the plunge.

In this respect, capitalizing on the knowledge economy requires an entirely new way of viewing the economic landscape. A cohort of knowledge workers and an emerging breed of knowledge entrepreneurs – who include entrepreneurial scholars turned intellectual venture capitalists – are set to respond to a new environment whose opportunities are created and shaped by initiatives among multiple empowered actors. Their combined energy and creativity is setting the scene for an entrepreneurial revolution that will transform that landscape.[2]

Exhibit 7: Knowledge Workers

Those who work with ideas, concepts, theories – that is, whose tool is a high-level abstraction and a symbol rather than a thing (Drucker 1993b: 170).

From the 1950s to the 1990s, from Peter Drucker to Robert Reich, knowledge workers and knowledge-intensive enterprises have been unceasingly taking over the reins in the most advanced economies.

In his book "The Practice of Management" (1954) Peter Drucker asserted: "In the United States… the class of employees that has been growing most rapidly in numbers and proportion is that of skilled and trained people." In the "The Age of Discontinuity" (1969) the same author affirmed that the "new industries differ from the traditional 'modern' industry in that they will employ predominantly knowledge workers rather than manual workers." Later in the 1990s he wrote in his "Management Challenges for the 21st Century" (1999): "[While the] most valuable assets of a 20th-century company were its production equipment [the] most valuable asset of a 21st-century institution, whether business or nonbusiness, will be its knowledge workers and their productivity."

Robert B. Reich, former US Secretary of Labor, interviewed by http://www.industryweek.com, has observed: "Globalisation has clearly shifted the balance of competitiveness toward human talents and skills… The core of the new enterprise in the 21st century will be talented people capable of quickly assimilating new knowledge and learning from one another" (Verespej 1999).

Exhibit 8: Knowledge Entrepreneurs

Knowledge entrepreneurs (KEs) are those who create value in the boundary-less knowledge markets through the infinite resource of knowledge that they put into action with the purpose of making advancements in the society, the economy and the environment.

(continued)

Exhibit 8 (continued)

Entrepreneurial Scholars

Entrepreneurial scholars, such as Marie Curie – an enterprising woman who became personally involved in the industrial application of her scientific results – show preference sets that are affected by the convergence of two character profiles:

- That of homo scientificus, breaking away from convention to search for ground-breaking discoveries, and
- That of homo economicus, with a special acumen for marketing and sales.

Source: Carayannis and Formica (2006)

The actions of KEs are influenced by their context. In fact, the deeper and broader the social interactions among KEs, the more value for innovative advantage can be assembled. This implies that "context gravitation" – that is, the force manifested by acceleration of KEs toward each other ("gravitation") in a setting of interrelated conditions ("context") – is at the core of any sustainability process that KEs trigger.

A highly charged "context gravitation" is linked to knowledge innovation zones (KIZs), which seek to coalesce their energy toward creating a uniquely human twenty-first century urban ecology driven by innovative advantages, and different kinds of knowledge hubs, including science & technology Parks (for the latter see Appendix 3).

A KIZ optimizes KE connections by getting knowledge to flow from the point of origin to the point of need or opportunity (Amidon and Bryan 2004). In this respect, KIZs are the epicentre of three forces that empower the innovation process –creativity, science, and advanced infostructures and infrastructures.

Exhibit 9: Knowledge Innovation Zones

A KIZ is a gravitation-sensitive context that takes the form of a geographic and conceptual space made up of complementary elements:

- Geographical: region, city, village
- Economic: sectors, clusters
- Social: communities of knowledge practice (see Part 1, Chap. 3).

Knowledge-based urban developments such as that of the Dubai Knowledge Village are cradles of KIZs populated by knowledge entrepreneurs and talented people, who become their elected citizens.

Exhibit 10: Dubai Knowledge Village

The Dubai Knowledge Village (KV) is committed to emerging as a global learning destination housing a wide diversity of international and regional higher education and training institutions. Since opening in October 2003, learning organizations with a stake in the region have steadily gravitated to KV, leading to its development into a 200-strong community that includes 15 international universities.

The new development at KV has immediately appealed to the student community, with the current intake due to reach 6,000 soon. The proportion of foreign students in the community has also risen from 15% to 25%. The presence of a growing multicultural student community has led to KV becoming a social microcosm in which diverse nationalities interact and build cultural bridges.

KV aims to build an international learning community that can promote innovation and entrepreneurship, supporting the development of ideas and talent.

Source: http://www.kv.ae/ – http://www.kv.ae/news/2003/21-07-2003.htm

The role played by knowledge entrepreneurs is one factor forging closer cooperation between academia and business in a knowledge economy. At the same time, it is necessary not to forget the crucial role played by the education system, which puts a drag on the potential for synergy. Teachers and researchers are slow in realizing the potential for creative and productive interface, today and for future generations.

The presence of substantial differences in such respects has influenced the achievements of past societies and will also influence what is to come in the future. An OECD report argued, with reference to the United States, that "stronger interactions between science and industry have characterised the innovation-led economic growth of the past decade and are currently helping the country to secure a lead in science-based industries ranging from IT and biotechnology to the new field of nanotechnologies" (OECD 2002).

By the same token, the UK has often been heralded as a star, for instance when it comes to its trade performance and its share of the workforce engaged in knowledge-intensive industries (Hutton 2007). A range of other postindustrial economies, such as Japan, Germany and France, have removed regulatory barriers and adopted institutional reforms to foster closer industry-science relations, while creating incentives for public research to join forces with business.

Some of the smaller advanced economies, such as the Nordics, Ireland, Australia and New Zealand, have been able to engineer more flexible and responsive institutions in these areas (OECD 2001). Many countries around the world are now looking for a way to make progress. Some still face difficulties in implementing reforms. Italy, for instance, is lagging behind in bringing academia and business closer together – when compared not only to the most advanced economies, but also some dynamic latecomers. Italy's ability to harness the right conditions for mutually reinforcing research and commercialization goals that can feed cutting-edge entrepreneurial opportunities is one of few available alternatives for boosting economic activity.

Exhibit 11: Main Attributes of Knowledge-Intensive Countries

- High level of entrepreneurial activity.[3]
- High-powered skills and competencies (in particular, interacting in socially heterogeneous groups[4]) for transferring knowledge from the point of origin (a laboratory) to the point of use (the marketplace) reinforce potential entrepreneurs in their decision-making.
- The mobilization of cognitive, creative and practical skills and competencies happens in a context of knowledge clustering that foments relationships among people for knowledge interchange conducive to higher productivity, economic growth and entrepreneurial activity from the exploitation of scientific and technological knowledge.

Natural-resource-based economies face significant challenges when it comes to laying the foundation for long-term sustained economic growth, diversification into high-value-added products, international competitiveness, and future social welfare. In order to compete successfully in the global market, and to continue to grow at a pace that will allow for progress in these respects, natural-resource-based economies must find ways to upgrade the skills of their labor forces, to strengthen enterprises' ability and willingness to innovate, and to lay the basis for mechanisms that can allow for expansion of new goods and services. A long-term growth strategy will require a strong commitment to the promotion of knowledge management, in key respects. The challenge is not primarily how to move gradually from the traditional raw-material-driven economy, but rather how to grasp and address the specific risks and opportunities that such an economy faces. This will require the materialization of a constructive interface between research and entrepreneurship, between knowledge creation and knowledge use, and hence for knowledge management in the private sector (see Part 1, Chap. 6) and public administration as well.

Resource-rich economies, such as those in the Central Asian and Gulf regions, are experiencing actions aimed at serving the needs of a sustainable society. These actions converge on projects of knowledge innovation zones as the most viable environment for nurturing, attracting and retaining three forms of entrepreneurship: green entrepreneurship ("ecopreneurship"), knowledge-driven business entrepreneurship and social entrepreneurship (see Part 2, Chap. 10).

Enabling the Knowledge Economy

Let us explore what is enabling the knowledge economy? The dramatic reductions in the cost of computing and communications driven by continued technological innovation and commercial exploitation of this innovation is creating the conditions for and infrastructures of a knowledge age.

At the core of the transition to the knowledge age, two so called "laws" encapsulate the key drivers which are enabling the transition from the Industrial age. Moore's law, originally postulated by Gordon Moore of Intel, has predicted and delivered computer power which on average is doubling every eighteen months at less or equivalent cost. To understand the magnitude of change that has happened in the performance/cost ratio of computing, consider if the same rate of improvement has applied to the auto industry. In this case it would be cheaper to buy a new Rolls-Royce everyday than to pay to park it overnight in a car park in a major city. Moore's law has become a competitive challenge for the global semiconductor industry acting as an innovation "beat" rate and focused innovation strategy, aligning and harnessing much of the innovation resources in the broader semiconductor ecosystem to enable repeated innovation and performance improvement.

The second "law" relates to communication and is named after George Gilder. It says that bandwidth grows at a rate three times faster than computer power which effectively means communication power doubles every six months. This is a result of continued technological innovation and continued broad diffusion of fiber and now wireless technologies. Indeed as wireless communication systems move to higher frequencies, bandwidth usability grows roughly with the square of the frequency shift (Gilder 2000). The emergence of wireless broadband technologies such as WiMAX make it economically viable to connect the next billion people to the Internet.

The continuing cost reduction in information search and transmission costs driven by Moore's and Gilder's Laws and the emergence of a semantic web, coupled with the accelerating connectivity of the human race to the Internet, is creating the conditions for a potential "Cambrian explosion" of knowledge. The European Internet Foundation envisages a unifying paradigm of the world of 2025 as one characterized by *mass collaboration*, enabled by the ubiquitous availability and use of high-speed, high-capacity digital communications networks, systems and services (Linton and Schuchhard 2009).

This scenario could be enabled by a series of multiplicative network effects (a network effect or network externality being the effect that a user of a service has on the value of that service to other people). Perhaps the best example of a network effect is the telephone where the more people have telephones, the more valuable the network becomes. In contrast to the telephone system and its primarily single usage model of enabling voice conversations (and also fax of course), the Internet enables many different voice, data and other services and can facilitate multiplicative network effects. One example of this is the case of YouTube which demonstrates a positive network effect where the more content that is generated and published by users, the more valuable the service becomes. In the case of YouTube there is also a negative network effect in that the high bandwidth required to transfer video leads to congestion and sometimes reduced quality of service (QoS).

Today's Internet is primarily based on an ingeniously simple but increasingly dated architecture where performance ultimately degrades as the number of users increases substantially. Fortunately there are architectures such as peer-to-peer computing architectures which behave oppositely in that performance actually improves as the number of users increases. In these scenarios, where more and more local copies of content are stored on user's computers, delivery time improves while traffic on the wide area network decreases significantly (Curley 2002; Curley 2004). Incorporation of such an approach in future Internet architecture designs could provide a more resilient and higher performing network to enable more seamless and faster knowledge transfer. Coupling ever improving network speed and access with ever improving kinds of community and team support technology such as advanced video conferencing, workflow and groupware will likely lead to reduced collaborative friction and improve both the creativity and productivity of group activity.

The Value of the Network

Underpinning the emergence of the knowledge age is the Internet which is a vast "network" with both centralized and distributed computer resources. At its core a network creates a series of "options" to enable a user to communicate, transact and collaborate. An "option" in this context similar to a financial option is the right but not the obligation to conduct an activity. David Sarnoff, a broadcaster pioneer is credited with "Sarnoff's Law" which states that the value of a network is proportional to the number of users attached to the network, for example for a television channel the more the viewers, the more attractive the television channel is to potential advertisers. Metcalfe's law named after Bob Metcalfe, inventor of Ethernet, states that the value of a network grows according to the square of the number of devices or people it connects. The scaling power of Metcalfe's law along with the astonishing performance of Moore's law has created an auto-catalyzing effect of driving global Internet connectivity, which is underpinned by a compelling economic logic based on scaling. The premise of Metcalfe's law is based on the number of transactions that can be enabled between connected users.

David Reed (2004) has pointed out that some network structures can create values that scale even faster than the square of the network users and this observation, which is now referred to as "Reed's Law," may become the dominant future value creation model in the knowledge age. A network which can enable the establishment of groups or communities that communicate and collaborate creates value exponentially according to Reed. Reed calls these kinds of communities Group Forming networks or GFNs and the viral expansion of services such as Wikipedia,

Facebook and Twitter are potential evidence of this exponential scaling. The dramatic expansion of Skype and its disruptive impact of lowering communication costs is also both an enabler and evidence of this phenomenon.

Future Internet: The Platform for Knowledge-Driven Entrepreneurship

"The Internet is the most important service and networking infrastructure providing the mechanisms for the digital society at large to function as an integrated entity" (Tselentis et al. 2009). With increasing connectivity and the advent of mobile broadband technologies such as WiMAX, the cost of attaching to the Internet continues to go down, while the ease of attaching to the Internet goes up. Technologies such as Satellite and WiMAX enable communities in difficult-to-reach areas to have access to the Internet.

Although designed more than 40 years, ago the simplicity of the Internet design and associated distributed governance has enabled the Internet to scale and become a critical infrastructure with much of the world's business and communication transacted over its ephemeral net. The core design of a network layer which enables dynamic path selection from a source of a packet to a destination (Tselentis et al. 2009) has enabled unfettered expansion of the network without this added scale adding significantly to the scale of the management overheads. In effect the Internet behaves as a complex adaptive system with distributed autonomy to enable efficient operation.

However, as the Internet becomes increasingly part of global society and business infrastructure, important weaknesses and vulnerabilities pose significant hazards in moving forward. For example packets are delivered on a best efforts basis, so it is near impossible to guarantee "quality of service." Services such as YouTube are consuming bandwidth and capacity at unprecedented rates while the emergence of an Internet of things, with many physical objects connected to the Internet via Radio Frequency Identity Tags (RFID) will mean that IP addresses under the current Internet Protocol V4 (IPV4) will run out soon.

The vulnerability of the Internet to hacker attacks is increasing each day with more sophisticated schemes being developed to undermine the operation of the Internet. Already, in 2000, Charles Wang of Computer Associates called a significant denial of service attack, which impacted many US businesses, the equivalent of an Electronic Pearl Harbor. Fortunately many people are aware of these issues and significant efforts to design a Future Internet which will avoid these issues and allow scaling to meet increasing demand are underway in a number of countries around the world. An example of this is the EU Future Internet Assembly (EU 2009), an expansive set of research projects with a goal of shaping the Internet of the Future. Such collaborative efforts can hopefully innovate and subsequently deliver a ubiquitous safe and high performance future Internet which supports and underpins knowledge-driven entrepreneurship.

Notes

[1] *Knowledge Innovation® is a registered trademark of ENTOVATION International Ltd. Permission was granted by the principals to use the term throughout this book.*
[2] Part 2, Chap. 8, provides an in-depth examination of both knowledge entrepreneurs and entrepreneurial scholars.
[3] According to van Stel, Carree and Thurik (2005), "... entrepreneurial activity affects economic growth, but... this effect depends upon the level of per capita income. This suggests that entrepreneurship plays a different role in countries in different stages of economic development."
[4] The OECD Project DeSeCo Definition and Selection of Competencies: Theoretical and Conceptual Foundations (http://www.deseco.admin.ch), which was led by Switzerland, defines three categories of key competencies: interacting in socially heterogeneous groups, acting autonomously, and using tools interactively.

Chapter 2
Foundation Laws of Knowledge Dynamics

Opportunities multiply as they are seized
(Sun Tzu)

Knowledge multiplies as it is used
(Amidon, Formica et al.)

The coevolution of ideas (the content) and their historical, social, organizational, and institutional forms (the context) has been the mainspring of economic growth throughout history.

It is the flow of energy that makes possible the coevolution that marks one civilization from the next. The energy circuit of the industrial age, such as that of sailing-ship or iron-hulled, ocean-going steamship, differs widely from today.

We live in the "knowledgefication" era in which the flow of energy is a flow of knowledge where innovation is instigated by the ability to create knowledge and manage it.

In a world awash with information, "knowledgefication" creates a difference in the sense that it allows people to discern from dozens of web pages the information needed and to gauge its relevance.

Knowledge is a Multiplier Asset to be Managed and Leveraged

Knowledge is bountiful provided that it operates as a multiplier – that is, the more it is shared, the more it grows, and the more it can be used innovatively. Otherwise, even human knowledge remains a scarce and costly asset.

The knowledge economy is seen as being fuelled by creative energy and a fountain of limitless resources because knowledge can grow exponentially as it is created and shared. "With knowledge, people can learn to do what they think. Intangible knowledge assets are reported becoming more valuable than tangible assets (Voice of the Entovation 100)."

The *First Law of Knowledge Dynamics* states that *knowledge multiplies when shared*. The resulting knowledge energy (or ken-ergy[1]), which triggers a process of

T. Andersson et al., *Knowledge-Driven Entrepreneurship*, Innovation, Technology, and Knowledge Management 102, DOI 10.1007/978-1-4419-1188-9_2,

organic growth, is manifested through a broad range of mechanisms. These include Innovation Management, Leadership for Value Creation, Knowledge Pattern Recognition, Knowledge Mapping, Knowledge Networks, Social Cybernetics, Mental Models, Situation-Handling, and Capital Systems. Since knowledge is inherently a human process, we must take care to optimize its creation and flow in ways that minimize loss in the transmission process (Amidon et al. 2005).

The general implication of the First Law is that a knowledge-based economy based primarily on intellectual, rather than financial capital, can lead to a more sustainable way to satisfy human wants and needs. To succeed, it demands an increasing amount of ken-ergy – again based upon human imagination and interaction – that can be progressively augmented, converted, and put in use to carry out the many tasks required without any loss in the processing system.

Exhibit 12: Intangible Knowledge Assets

Unseen assets: An ill-defined category that embraces research and development projects, design patents, brands, in-house training, high skilled staff, company-specific software, and leadership.

Firms must adopt "invisible balance sheets" of intangible assets with flows of knowledge in between.

Intangible investments show a remarkable upward trend. In 1970, investment in these so-called intangibles was a mere 40% of investment in tangible assets: now it is 125%.

Source: Andrew Hill, Financial Times, Monday August 6, 2001: 13; Hutton 2007. See also Appendix 4.

Exhibit 13: Organic Growth

Organic growth is a self-organized, self-sustaining, and self-reinforcing formation of interconnected businesses, whose seed is a "food" molecule (catalyst) without which the "business reaction" would proceed only with a great difficulty. High quality local resources such as skilled individuals and local-rooted entrepreneurial heroes usually act as catalysts.

Exhibit 14: Intellectual Capital

Intellectual capital is the ability to transform knowledge and intangible assets into wealth creating resources.

The classic categorisation of intellectual capital includes relationship, structural and human capital.

Structural capital, broken down into customer capital (i.e., penetration, coverage, loyalty, and profitability of the customer) and organizational capital,

(continued)

Exhibit 14 (continued)

is defined as the organizational capabilities of the organization to meet market requirements.

Human capital is the capabilities of the individuals required to provide solutions to customers: for example, light, not light bulbs; transport, not cars; nutrition, not foods; experience, not media. A perspective very much akin to Thomas Edison's business model whereby "success is achieved by selling to customers... the fewest number of light bulbs necessary to supply them with the light they wanted" (Wyllie 2002).

An organization's intellectual capital (IC) consists of the intangible sources of value related to employees' capabilities, the organizations' resources, and ways of operating and the relationships with its stakeholders. Examples of IC include employee competencies and motivation, relationships with stakeholders, management structures and processes, and patent and other intangible property. An organization's success is significantly affected by how well it can utilize IC, for example, in creating new innovations. Thus, IC management has emerged as an important practice and an area of research.

Innovation is the Process that Converts Knowledge into Action

The Second Law of Knowledge Dynamics states that *value is created when knowledge moves from its point of origin to the point of need or opportunity*. This affirms that the real benefit of knowledge lies in action; and that innovation is the process whereby knowledge is put into motion or used. This process of innovating knowledge requires high-powered knowledge energy flows (see First Law) supported by wide bandwidth connectivity and rich interrelated actions. The Second Law emphasises effectiveness whereby actions aiming at "doing the right thing" (effectiveness) prevail over those addressed at "doing things right" (efficiency).

Two corollaries descend from the Second Law. The first is that being efficient without being effective does not necessarily lead to the desired result. This brings about a stress in the efficiency process that can imperil the minimisation of effort and cost in the next round of resource utilisation. The second points out the simplicity of that mechanism for transmitting ken-ergy, which takes the form of voluntary relationships experimented by free communities of knowledge practice. These relationships contrast sharply in interests and abilities with the method based on permission enforced by law (i.e., hierarchical and command-and-control) – the latter portending a planned and corporatist society where the chokehold of vested interests plays a relevant part in the paralysis of the ken-ergy transmission mechanism. This paralysis is attributable to a primary role conferred on planners as policymakers (Amidon et al. 2005).

Exhibit 15: Knowledge in Action

Knowledge can be converted or embedded into processes, goods, and services. Different sources of knowledge can join together to increase the value of a product or service. Knowledge can be "productised" and then commercialised in the marketplace.

Today's challenge is shown in Fig. 1, which shows three hills of self-organized knowledge. The first hill represents the knowledge domain of the pre-Industrial Revolution age. During that time, there were local communities that reached the top of the hill (P1) through relentless innovation in the manufacturing system.

Groundbreaking innovations in mechanical engineering that permanently altered market conditions led to the decline (point 1 at the left side of Fig. 1) of the manufacturing system. Crossing the area of ignorance (point 2) in order to savor the benefits of the incoming Industrial Revolution was a long and painful run, ending with the conquest of a new peak (P2) – that of the Industrial Age. This was the ability to combine theoretical knowledge with manual practice at innovative institutions for technical education and of an impressive, widespread learning of how tangible assets ought to be handled – thanks to a range of spin-offs acting as teaching companies.

Today, the acquisition of new theories and modes of organization is becoming more and more important as innovation accelerates under the stimuli of the knowledge economy. "Let go at the top" of a new peak (P3) involves four key factors.

Namely:

- A knowledge process for mapping and exploring the holes of ignorance
- The capability of managing not only tangible assets but also intellectual capital
- The access of local communities to entrepreneurial higher education institutions in coopetition with traditional academia and technical schools
- An increasing number of knowledge-based, knowledge-driven start-ups from the research and learning environment – the very engine of firm creation in the digital era as the shop floor was in the industrial one.

Value Creation – Through Knowledge Innovation – as a Collaborative, Open Process

The Third Law of Knowledge Dynamics states that *mutual leverage provides the optimal utilisation of resources both tangible and intangible. It* asserts that collaboration and the value of leveraging the knowledge of one another create a greater wealth and sustainability for us all. Unlike vested interests playing against

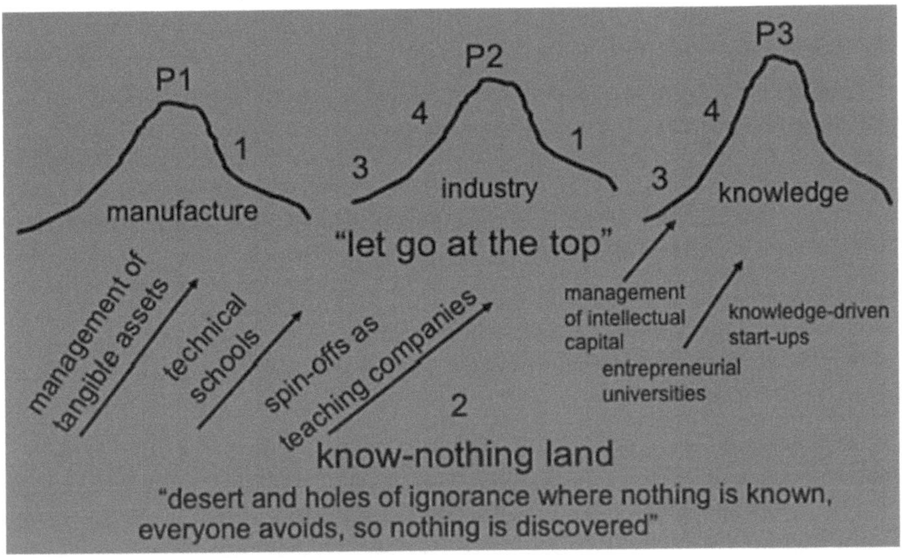

Fig. 1 Hills of self-organized knowledge

competition, collaborative efforts made by those agents who put knowledge into action are incentives not to collude but to combine cooperation and competition so as to enhance precompetitive forces working for the general interest of the knowledge society. There are multiplier implications that operate within and across network boundaries. Synergy and symbiosis are natural outcomes of the human interaction in ways that provide profound network effects (Amidon et al. 2005).

Players in the cooperative environment thrive as they accept that whenever there is a trade-off between short-term profits and investment for breeding the competitiveness, the latter has to prevail. As their motto goes, "a larger cake is better than a larger slice" (Kay 1995).

The distinctive feature of cooperation is the predominance of horizontal and dynamic relationships from which a complex balance of partnership and rivalry

Exhibit 16: Forms of Collaboration and Network

Collaboration

Collaboration is an enduring process of learning how to manage relationships. Trust and social capital (see Exhibit 36) are the "genes" that dictate the evolution of such a complex, adaptive web as collaboration is.

"The quality and quantity of meaningful collaboration often depends upon the tools used to create it… Collaboration is like romance… it can't be routine and predictable" (Schrange 1990).

(continued)

Exhibit 16 (continued)

The primitive form of collaboration is defined as "coordination". Coordination enhances the process of "simultaneous adoption of identical or complementary strategies by independent agents" (Kay 1995) whose relationships are informal and implicit, based on unwritten rules and unwritten codes of behavior.

Distinctive features of the relationships prevailing in the coordination environment are the following:

- Arm's-length and kinship/family ties
- Informal, opportunistic, and short-term relations
- Spontaneous dissemination of implicit knowledge
- Joint problem solving

"Cooperation" is a form of collaboration that allows the parties to maximize the joint product of their relationships rather than the individual returns.

The cooperative pattern embodies a good deal of competition within (and not only outside) the relationship for both the production and distribution of the net gain of a non-zero-sum game. This results in a variety of situations of competitive cooperation or coopetition across parties.

As Nalebuff and Brandenburger explain: "[There] is cooperation when it comes to creating a pie and competition when it comes to dividing it up. In other words, [there] is War and Peace. But it's not Tolstoy – endless cycles of war followed by peace followed by war. It's simultaneously war and peace... You have to compete and co-operate at the same time. The combination [that is, 'co-opetition'] makes for a more dynamic relationship than the words competition and co-operation suggest individually. [In the co-opetition game] your success doesn't require others to fail – there can be multiple winners" (Nalebuff and Branderburger 1996: 4–5).

Cooperation and competition mix in various ways. Hampden-Turner and Trompenaars (1997) state that varying blends of competing and cooperating correspond to different coopetition games. The two authors call "finite games" those situations in which cooperative competing is an attitude that leans toward competitiveness. "In finite games you compete, the better to cooperate. Conversely, endless or 'infinite' are those games that incline toward cooperativeness. In infinite games you cooperate better to compete" (Hampden-Turner and Trompenaars 1997: 25; Fig. 2).

"Coopetition" is perceived as a two-tier game: before "domestic" (inside the relationship) cooperation across the parties and after "foreign" (outside the relationship) competition between them. Its intrinsic nature is the unification of, not the division between, cooperation and competition. This is equivalent to saying that coopetition fits the principle of multivalence or fuzziness instead of that of bivalence or two-valuedness (Kosko 1994). It means a large spectrum of competition and cooperation options. So, for example, to some degree the area of cooperation might be related to the "untraded" factors of the interdependency,

(continued)

Exhibit 16 (continued)

for which collaborative actions are developed by autonomous (ownership-independent) teams, whereas to some degree the "traded" factors might provoke competition across parties inside the relationship long before those outside the relationship. In short, coopetition is an overlapping world made up of competing and cooperating teams whose borderlines are "domestic" cooperation and "foreign" competition.

Network

Internal Network – The networker creates an internal market. The network partners, who are the networker's collaborators, interact between them and with the networker through the open market price mechanism. The networker plays a central role, standing out in terms of experience, strategic capability, and strength of personality.

Stable Network – Independent, specialised partners are connected to a "core unit" through long-term relationships and contractual solutions. Each independent partner keeps and develops its distinctive competence, even serving other players outside the network. The core unit stands out in terms of experience, strategic capability, and strength of personality.

Dynamic Network – This is based on short-term or one-time relationships. Independent, specialized partners are coupled contractually for a particular project and then part company for the start-up of a new venture, each of them operating at a specific segment of the value chain. There are no special, privileged relationships. A central unit of coordination plays a role primarily administrative (Miles and Snow 1992).

Network Code

Voluntarism	– Partners are free to withdraw from relationships they believe are unfairly structured.
Openness	– Network relationships are external and thus highly visible to all parties.
Explicitness	– External, visible relationships tend to be explicit.
Simplicity	– "The less you sign, the more you achieve."
Performanceness	– Network relationships are guided by performance rather than by procedures.
Accessibility	– Full-disclosure information systems ensure that all decisions are made objectively and fairly.
Self-renewal	– Many autonomous, intelligent work units interacting rapidly with the outside world and each other, and quickly rearranging themselves to solve new problems.

Source: Adapted from Miles and Snow (1992)

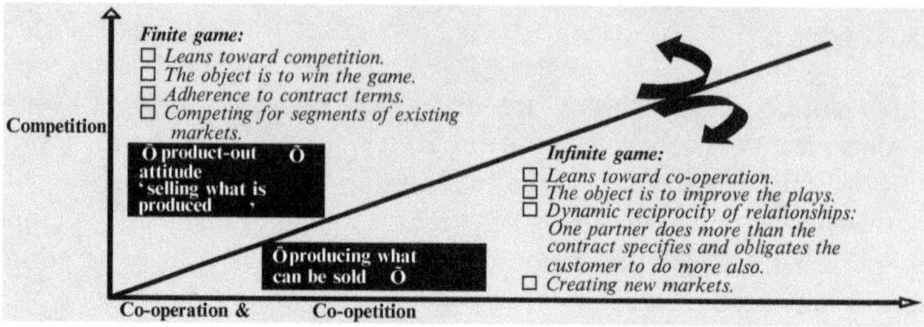

Fig. 2 Competition, cooperation, and coopetition: finite and infinite games. Source: Adapted from Hampden-Turner and Trompenaars (1997)

results in a permanent state of disequilibrium. Cooperation is needed to increase the dimension of the market, while competition remains the essential ingredient that motivates players to strive for excellence. In contrast, the traditional competitive paradigm can only lead to each player to struggle for making its own share of the market cake increasing more rapidly than the cake shrinks. It is an unbearable effort in the long run. In fact, this strategy requires aggressive pricing and/or costly product customisation. It leads to lower margins and diminishes the level of resources available for the development of strategic competencies on which to build competitive advantages.

Through cooperation, companies learn how to practise the very sophisticated "coopetition" game – the frontier of collaborative practices. Coopetition is to compete and cooperate with the same players.

Coopetition is encouraged by web-based trading applications. Within an online community tailored to serve a specific industry, activities can be coordinated in such a way that they are good for the entire industry. A "net" infostructure appears more promising than a "pit" architecture where the holders of information behave as the owners of wells. The available information is that the well's owner wants supply. Partial and opportunistic disclosure of information engenders a climate of distrust within the customer–supplier relationships. Conversely, while maintaining the dynamics of competition, the online enriched flow of information allows companies to improve the coordination of activities so that, for example, it comes more naturally to sell inventory in excess to direct rivals. Revealing inventory, pricing, design specification, and other kinds of hot information to competitors is something that sounds strange in the conventional business culture. Nevertheless, this conduct becomes a crucial challenge, albeit with unpredictable consequences, as soon as companies are aware that they have to change their own behavior going online (Henig 2000).

Note

[1] Ken means perception, understanding, range of vision, view, sight; and to make known, to know (a person or thing), to recognize, to descry (i.e., discern something difficult to catch sight of; discover through careful observation or investigation; to have an understanding of something). In other words, ken captures the essence of having knowledge AND having a sufficient range of vision to know what to do with it knowledge (i.e., the energy of knowledge). See Debra Amidon, *Innovation Strategy for the Knowledge Economy: The Ken Awakening*, Butterworth-Heinemann, 1997, pp. xxii.

Chapter 3
The Knowledge Value Chain

It is no good to try to stop knowledge from going forward.
(Enrico Fermi)

The sustained phase of transition to economies characterized by considerable, and sometimes revolutionary, advances in science, technology, and related industries, coupled with subsequent profound changes in economy and society, has increased the importance of the knowledge-intensive phases of production for value creation. As enterprises become more reliant on technology, they will become more dependent on knowledge.

Accordingly, policy makers in a growing number of countries have become increasingly concerned with the management of the entire knowledge chain: from creation to transfer (i.e., the diffusion, conversion, and entrepreneurial exploitation of scientific and technological knowledge). The knowledge chain also has profound implications for higher education institutions and business schools, which, to be successful, need to help companies create knowledge and become part of knowledge streams.

Exhibit 17: Knowledge Management

Knowledge management is about the flow of meaning. It refers to social communication processes supported by collaboration technologies (e.g., the Internet, Intranets, wireless devices) to create new knowledge and to efficiently reuse knowledge to obtain and maintain competitive advantage through improved performance and decision-making of workers.

Knowledge management is wrongly assumed as a simple extension of information management. In fact, the former entails major changes not only to technology but also to corporate culture, operations, and relationships.

Whichever way one looks at knowledge management, its primary issue is that of people management (see Appendix 5).

T. Andersson et al., *Knowledge-Driven Entrepreneurship*, Innovation, Technology, and Knowledge Management 102, DOI 10.1007/978-1-4419-1188-9_3,

Knowledge Creation

Knowledge creation is the process by which new energies are channeled toward the development of successive layers of innovation, each creating a new tier of business. The community affected by far-reaching structural changes is transformed.

Tacit or intuitive (as opposed to explicit) knowledge is critical to knowledge creation. The seminal work of Nonaka and Takeuchi (1995: 56) assigns to knowledge creation a key role in the mobilization and conversion of tacit knowledge. Nonaka and Takeuchi's knowledge-conversion process is shown in Fig. 3.

Creating knowledge needs time for reflection. And the fact of sharing helps knowledge creation. At Unilever, the head of knowledge management has defined as "organizational alignment" as the observation that "the more an organization is connected, the more it can combine insights and knowledge to get creative breakthroughs". Unilever's formula is "alignment 50%, processes 30%, and IT 20%" (Jackson 1998). All in all, in a networked life innovations happen faster and errors are spotted sooner. This seems to be true in business as well as in the academic arena.[1]

Knowledge creation lays the ground for the fastest expanding industries of the future that would not have existed otherwise. In such an economy, diversity rather than specialization holds the key to long-run success.

Knowledge creation differs from knowledge recycling and diversion. Knowledge recycling and diversion is the process by which existing resources are recycled for the purpose of protecting established practices within a given craft or industry, and new energies are diverted toward the same goal.

Industrial Era companies are used to making objects. Hence, their trading activity is mainly one of the exporting and importing atoms (i.e., physical objects). Producing more hardware than software, more machines and apparel than information and knowledge, companies hang on the efficiency and effectiveness of their carriers to deliver atoms (Negroponte 1995).

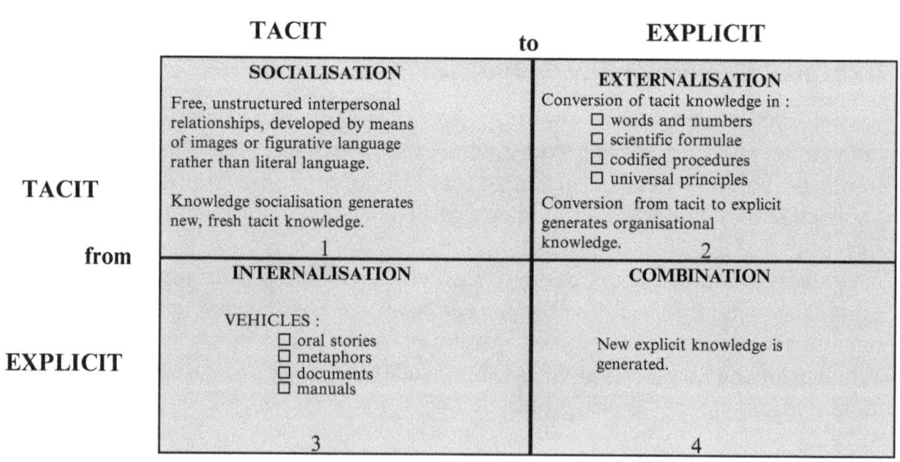

Fig. 3 The knowledge-conversion process

Compared with physical products, the exploratory phase of knowledge creation – think about the design and tooling of a new chip, a new drug, or a new gene therapy – is very expensive because it needs a big initial investment, but the production cost of copy in terms of knowledge-intensive products or services becomes cheaper and cheaper, almost nil. Therefore, knowledge-based businesses benefit from increasing returns, or decreasing costs. As the market grows, returns can increase enormously and producers have the power to charge high enough prices to pay for the exploratory phase (Klein 1996).

Exhibit 18: Tacit and Explicit Knowledge

Tacit Knowledge

The informal, occupational wisdom and experience generated by people grappling with everyday problems and passed on in café-type communities and online communities. One can gain tacit knowledge orally, by direct experience, and by trial and error.

Tacit knowledge is a set of intuitions, insights, hunches that are gained through the use of metaphors, pictures, or experiences. It is subjective, not easily perceivable and expressible, below the level of awareness and is therefore difficult to communicate (Nonaka and Takeuchi 1995).

Explicit or Institutionalised Knowledge

The official rules are codified and written down in books and manuals.

Explicit knowledge is formal and systematic, articulated in words and numbers, and found in manuals, books, databases, and files (Nonaka and Takeuchi 1995).

Exhibit 19: Increasing Returns

Knowledge is a near-endless resource market, not subdued to the iron law of scarcity and diminishing returns.

"The more stuff we know, the greater the base of existing human knowledge, and the greater the payoff from the next discovery. Knowledge is what economists refer to as an increasing returns phenomenon...in the eighteenth century Jacques Turgot showed that most production processes exhibit the opposite quality of decreasing returns. For most types of production processes, whether it is farming, manufacturing, or services, as one inputs more and more resources, the marginal returns get smaller and smaller... [vice versa] the more we invest in knowledge cumulatively over time, the higher the payoffs. An hour of R&D invested in microchips and biotech today has a higher payoff than an hour of R&D invested in steam locomotives and telegraphs in 1900" (Beinhocker 2006).

(continued)

Exhibit 19 (continued)

The most striking examples of benefits reaped from increasing returns are at the interface of industries such as personal computers, consumer electronics, information, telecom, music and films, and education. This statement by Sony's chairman is a case in point.

Nobuyuki Idei (2000), chairman of Sony, says:

"The law governing the manufacturing industry is a law of diminishing returns. That is, at some point it costs more to produce and sell the next widget than you can justify economically."

"But technology has created exceptions. The most obvious is digital content such as software, where the marginal cost of producing the next copy of the product is near zero. That phenomenon brings about a process known as increasing returns. Microsoft has most famously proved this to be true, but it applies to any popular software platform."

"There's also an exception in the manufacturing business. With semiconductors, including memory chips and microprocessors, the cost of creating a new chip design and factory is enormous but the cost of making each chip after the first is small."

Additionally, Kurzweil (2001) in an analysis of history conclude that technological progress follows an exponential growth pattern and he predicts "technological change so rapid and profound that it represents a rupture in the fabric of human history."

Communities of Knowledge Practice

Knowledge creation, shared knowledge, and collective intelligence are fundamental precepts of the knowledge economy. These precepts are encouraged, on the one hand, by technological factors such as the steady expansion of the information and communication technologies, and, on the other hand, by human factors such as the formation of "communities of knowledge practice," most of them in the form of online knowledge communities, which are part of cross-border and cross-cultural knowledge networks (see Part one, Chap. 4).

Exhibit 20: Collective Intelligence

Collective intelligence is a knowledge asset that stems from the primary process of reason performed by each individual alone.

Industrial capitalists, who act as investors in plant, machinery, and equipment, add to the stock of physical capital. Likewise, knowledge capitalists invest in relationships that contribute to the formation of intangible capital. By doing so, knowledge capitalists create a community context conducive to a collective

(continued)

Exhibit 20 (continued)

agreement on specific contents through spontaneous interactions of many, heterogeneous individual thoughts.

We call "collective intelligence" this sort of compromise, which is non-contractual and experimental, and is designed to maximize the creative collaboration of small groups of people addressing issues that interest and concern them all.

Ayn Rand has been a strong opponent of the collective intelligence concept. As the novelist–philosopher put it in her bestseller, "The Fountainhead": "…the mind is an attribute of the individual. There is no such thing as a collective brain. There is no such thing as a collective thought" (Rand 1943). Neither it is a byword for collective intelligence.[4]

Exhibit 21: Communities of Knowledge Practice

Xerox's John Seely Brown and Etienne Wenger have defined communities of knowledge practice (CKP) as "groups of people who've worked together over a period of time... not a team, not a task force, not necessarily an authorised or identified group...They are peers in the execution of 'real work'. What holds them together is a common sense of purpose, a real need to know what the other knows... a shared passion for something that they know how to do and to interact regularly to learn how to do it better."

Communities of knowledge practice attract all sorts of characters, unlike communities of practice that ties membership firmly to job description (e.g., the company-based communities of repairmen in the packaging machinery cluster – Bologna, Italy; see: Formica 2003a; Chap. 1).

"CKP harnesses the creativity and promote cross-fertilisation of ideas necessary for prosperous innovation. Innovation becomes the glue that bonds together diverse constituencies" (Amidon 1997).

"CKP can't be created or designed, but can only be discovered and supported. The most exciting communities exist at the borders of organisations, attracting all sorts of characters" (Tom Peters).

A contemporary example of such a community is the Innovation Value Institute (http://www.ivi.ie), which is a community of IT executives and professionals from many companies working together with some academics to pool knowledge to build tools to solve problems in IT management which are too large and difficult to allow any one company or organizations to solve by themselves. Collective Intelligence is pooled using common blueprints, based on a draft periodic table of business processes which CIOs need to manage to create value for their organization (Curley 2007).

(continued)

Exhibit 21 (continued)

The Innovation Value Institute (IVI) at the National University of Ireland, Maynooth (NUIM) is an excellent example of the power of collective intelligence leveraged through an open innovation process.

The IVI was established by Intel and NUIM to help drive a structural change in the way companies and governments get value from IT. Its key output, a new business tool called the IT Capability Maturity Framework (IT-CMF), was originally developed by Intel and further developed by a consortium of 40 companies and organizations including Boston Consulting Group, Microsoft, Ernst & Young, Axa, Chevron, SAP and BP. The IT-CMF helps deliver greater business value from IT while reducing the complexity of IT choices for CIOs.

The open innovation and not-for-profit configuration of the IVI consortium has enabled the technology ecosystem and enterprise end-users tackle a problem, which is larger than any one company could solve on its own, that of how to systematically improve the return from information technology investments. The open innovation approach used where IT executives work together to create best practice tools coupled with academic input and oversight has enabled delivery of practical value add tools far faster and cheaper than any one company could have achieved on their own.

In the knowledge economy, communities of knowledge practice rise to the challenge of playing the role of the Industrial Age cafés. In Italy, blue-collar workers and technicians were used to meeting in cafés where, playing cards at small tables, they engaged passionately with each other in discussions on technical advancements and the new organisational forms that could be adopted in their companies. Interactions in an informal and open relational space of conversation gave birth to new companies in new market niches whose founders were the participants in those lively but nonconfrontational talks. Thus, for example, playing cards in cafés instead of playing by business cards in formal debates was the source of a stream of companies and technical innovations that built the renowned cluster of packaging machinery in Bologna (Formica 2003a, Chap. 1) and motor racing and sport cars near Modena, Italy (Formica 2003b).

Exhibit 22: Café-Type Communities

Café-type communities of practice are life forms, whose behavior is organized from the bottom up.

Café-type face-to-face-based communities are suitable for enabling participants to exchange and transfer "skills" (manual knowledge) or "technical elements of tacit knowledge."

The odd point is that face-to-face interaction induces "conformity effects" and "group thinking," which encourages participants to think inwards.

Source: Yamazaki (2004)

As Yamazaki's argument suggests, the work of online knowledge communities is to generate ideas that will be converted into new ventures for the knowledge economy.

Exhibit 23: Online Knowledge Communities

One feature of these communities is the sense of individualism felt by their participants, who "behave as self-contained decision-makers" instead of going along with the "group-type behavior" path seen in face-to-face-based communities.

Another feature is that most knowledge community-participants tend to use nicknames. The use of nicknames makes interaction easier because knowledge exchange takes place on an equal footing (in low context communities, that is, irrespective of status considerations).

Insofar as these features prevent "conformity effects" from occurring, online knowledge-community participants would settle for being schooled in the art of outward looking. As a result, new companies and businesses would be established that look forward rather than backward. In other words, there would be more start-ups whose scope extends well beyond the horizon of the traditional industrial basis to envisage the needs of the knowledge economy.

Exhibit 24: Yamazaki's Dual-Context Society

Low Context Communities

- Member participation is very vague and wide open
- Participants are on an equal footing
- Strength of weak ties
- Non face-to-face interactions in online-based knowledge communities
- Exchange of cognitive elements of tacit knowledge
- Self-supporting people who think forward and are outward-looking
- Unconventional effects

High Context Communities

- Job description fixes the limit of participation
- Seniority rules prevail
- Strong ties that bind
- Face-to-face interactions
- Skills transfer and exchanges of technical elements of tacit knowledge
- Inwards and backwards group thinking
- Conformity effects
- The introverted nature of high context communities means that they can easily be held hostage to the sway of vested interest groups for which the protection of their members becomes a life and death matter

Source: Adapted from Yamazaki (2004)

Communities of knowledge practice aim to produce new routines breaking the old ones. This requires the ability and willpower to constantly upgrade, deploy, and direct their efforts toward business processes at the threshold of the significance where the importance of a new round of innovation precedes its momentum (see Fig. 4).

In this respect, communities of knowledge practice exhibit an ever-increasing tension between the experience curve, which brings to the capitalisation of the existing structure of knowledge, information and behavior, and the learning curve, which evokes the need to unlearn or discard knowledge. The former curve boils down to the capitalisation of the existing knowledge, information, and behaviour for which there is a need of predictability (no surprises – i.e., "low information entropy"). The latter is about communicating unexpected news (unexpected news is welcome – i.e., "high information entropy"[2]) (Gilder 2002: 104–107).

Exhibit 25: Experience Curve and Learning Curve

The experience curve describes the decrease in cost (increase in efficiency) due to continuous changes, experience, and scale in the manufacturing of any product.

The learning curve attempts to measure the increase in productivity induced by discontinuous changes (or "innovativeness").

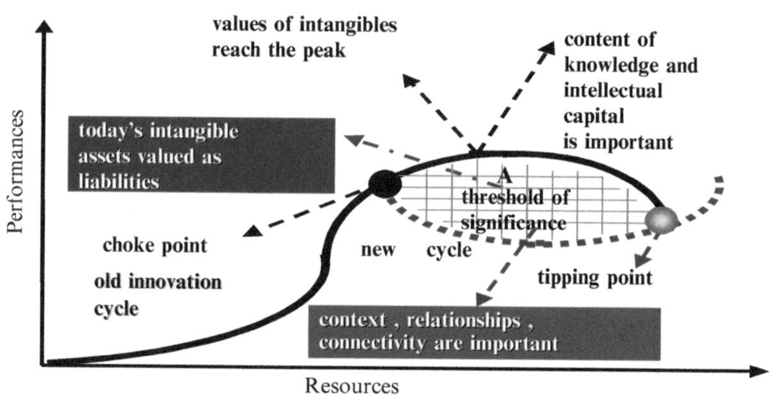

Fig. 4 How communities of knowledge practice work at the threshold of significance

The threshold of significance coincides with the "strait of discontinuity" (region A in Fig. 4) between two innovation cycles where past and future coexist in the present. In Fig. 4, the moment when unexpected, radical changes suddenly become a reality is represented by a grey dot at the exit of the "strait of discontinuity." This is the moment of critical mass or "tipping point" at which the significance of a new round of innovation precedes its momentum. As Gladwell (2000) puts it: "the world of the Tipping Point is a place where the unexpected becomes expected, where radical change is more than a possibility: it is – contrary to our expectations – a certainty."

Collective actions by participants in communities of knowledge practice enforce courses of action that consist first in detecting and then taking possession of particular business activities that control the flow of profits throughout an industry (so called "choke points"[3] – the black dot in Fig. 4) at the entrance of the strait, so as to open up the navigation to start-ups in technologically progressive new firms enjoying fast sustained growth.

There is an inevitable time lag between the decline of today's stars and the success of tomorrow's businesses. Nontangible assets show the highest values when the economy has reached the peak of the old innovation cycle. But this makes people blind to novelty.

A tiny vanguard of individuals, who are giving birth to far-reaching changes to come, is overshadowed by the majority made contemptuous by the current values of human capital, marketing skills, patents and brands, and so on – when all these intangible assets are indeed incoming liabilities. So, a stubborn complaisance toward the present endowment of intangibles is an impediment to looking outward from today's business environment and forward into the future. At the surface it seems that the economy has been doing everything right, but seen from a closer perspective that mentality is counter-productive and condemning the country to its future decline.

By challenging the prevailing thinking that perceives the seeds of future prosperity in the content of intangibles, communities of knowledge practice address their efforts to the context in which unseen assets ought to be embodied.

Cross-Boundaries and Cross-Rival Communities of Knowledge Practice: Their Evolution into Superior Forms

Throughout history, individuals and businesses have clustered into networks. The medieval craft guilds, which emerged from increased manufacturing specialization, were an early incarnation of the modern network.

An innovative mind is a knowledge worker's most important means of production. Knowledge workers need to collaborate and communicate to create value and change real wealth into the next good idea. Networking produces a free exchange of ideas and allows for concepts and connecting theories, methods and techniques – which on their own would be inadequate – to be shared.[5]

Communities of knowledge practice tend to be cross-boundary in geographical, cultural, and functional terms. They may even include people from rival companies. Investments in digital technologies are stimulated as technical vehicles to orchestrate a continuous knowledge and learning environment.

Exhibit 26: Past and Present Examples of Cross-Boundaries and Cross-Rival Communities

The Lunar Society was a group of visionary, multitalented, nonconformist, nonacademic practical types in Birmingham in the mid-eighteenth century. They were driven by curiosity in the workings of the natural world. Lunar Society members were responsible for a wave of innovation set in motion by the discovery of oxygen (Joseph Priestley), the fine-tuning of the steam engine (James Watt) and the modern commercialization of pottery (Josiah Wedgwood). Their achievements include fossil classification, telescope manufacture, and the creation of sparks of electricity (Uglow 2002).

The seventeenth-century geographer, Father Vincenzo Coronelli of the Order of Friars Minor in Venice, was the catalyst of a community of practice, the Cosmographic Academy of the Argonauts. This group included princes, merchant-politicians, and illustrious savants from all over Europe. Thanks to geographic information obtained inside the community, Father Coronelli improved his cartographic and printing workshops in the Franciscan convent in Venice (Wills 2001: 9–11).

Entovation E100 is a social, business intelligence network for global learning formed by 100+ theorists and practitioners representing around 50 countries and including thought leaders, CEOs and senior managers, government officials and academic researchers. Entovation E100 members include experts in performance measures, competitive analysis and alliance strategy, as well as in computer/communications technology (Amidon 2002; Chap. 13).

Participants in the "e-ducation Without Borders" conference (http://ewb2007.hct.ac.ae/ewb2007/), which is held in the United Arab Emirates every 2 years, are students and experts from unconventional backgrounds as well as from more traditional disciplines. They are mixed together in such a way that each player comes to terms with the strengths and weaknesses of the others. Diverse conversations between people of different cultural and social background facilitate the formation of new ideas and prevent the sharing of the same knowledge.

Asian communities stimulate horizontal thinking by forming a community of competitors. "In the semiconductor industry, Yin and Lin (2002: 10–11) observed that competitive advantage depends on the ability to react quickly to markets and to plan research and development. Forming a community of practice with companies in the industry enables them to keep up with constant change in the marketplace. However, organizational commitment and intellectual property have become major concerns in building communities with competitors."

Organisational Knowledge Creation

From the standpoint of knowledge produced by the scientific community, creation of new knowledge through research, often collaborative in nature, includes knowledge presented in scientific papers, pieces of scientific knowledge not yet formalized in a scientific paper and that type of tacit knowledge concerned with the methodology used in scientific processes and experiments.

Knowledge creation and organisational knowledge creation are the means of "breaking away [companies] from the past and moving them into new and untried territories of opportunity" (Nonaka and Takeuchi 1995: 4).

People who coalesce in communities of knowledge practice become key players in generating a new kind of business organisation: the "knowledge-creating company." A knowledge-creating company adopts ways of working traditionally associated with education. By absorbing intellectual capital and thereby moving its centre of mass from material to intangible processes, the knowledge-creating company (KCC) resembles a training and research institute, which learns and encourages learning in its people (Handy 1995), but of a generative rather than an adaptive type.

In the words of Nonaka and Takeuchi (1995: 44), the first kind of learning "is establishing new premises (i.e., paradigms, schemata, mental tools, or perspectives) to override the existing ones," which are influenced by past experience.

Exhibit 27: Learning

Learning is the increase in the amount of response rules that are the result either of existing premises or of new ones.

Generative learning increases the amount of proactive responses as opposed to reactive responses.

Because an organisation cannot create knowledge on its own, KCCs encourage the initiative of their people and provide a shared context (say, a community of knowledge practice) in which individuals can interact with each other.

"Team members," Nonaka and Takeuchi argue, "create new points of view through dialogue and discussion. This dialogue can involve considerable conflict and disagreement, but it is precisely such conflict that pushes employees to question existing premises and to make sense of their experience in a new way. This kind of interaction facilitates the transformation of personal knowledge into organisational knowledge." This is new knowledge that KCCs employ to reconfigure their business relationships according to market needs and so as to develop commercially valuable innovations.

The process of transforming personal knowledge into organisational knowledge fails if it takes a course of action for which its core audience reflects the interests of existing businesses and no longer the inner emotions and enthusiasm of the individuals per se. Because people (not companies) possess knowledge, the consequence of

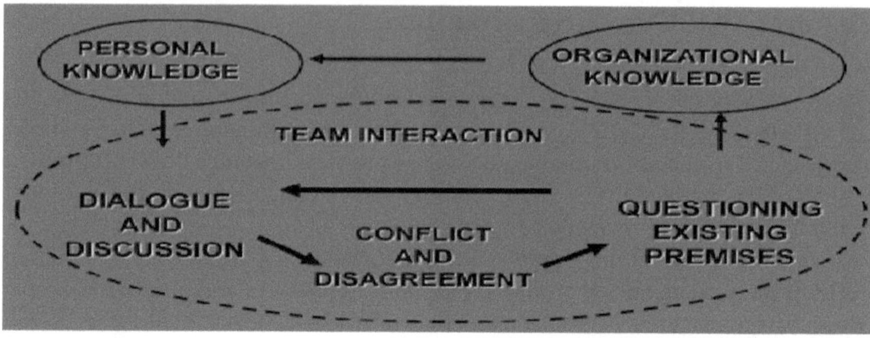

Fig. 5 Transformation of personal into organizational knowledge Source: Adapted from Nonaka and Takeuchi (1995)

Exhibit 28: Organisational Knowledge Creation

"Organisational knowledge creation is the [generative-] learning companies' ability to create new knowledge through systems, routines, data within an organisation which are only imperfectly understood by any individual member" (Kay 1995).

giving a much greater voice to companies is that immense care is devoted to passing information around and between organisations instead of creating and disseminating new knowledge.

In the light of the above, the KCC's work environment is a living organism rather than a machine for processing information, within which intensive and laborious interactions among team members allow each individual to learn from others as well as to acquire knowledge from outside. In such a way knowledge is created, and organisational knowledge creation stems from the conversion of tacit into explicit knowledge (Nonaka and Takeuchi 1995; see Fig. 5).

Knowledge Transfer

Knowledge transfer (KT) is the process that puts knowledge into action. It can be seen as a basic process for speeding up the flow of knowledge among all components of the knowledge chain. To the advocates of KT, this process brings about a self-reinforcing circuit between productivity, economic growth, and entrepreneurial activity.

Knowledge in action that secures availability of pertinent knowledge at the point-of-action, and just in time, has the power to produce innovation as its desired result (Wiig 2005). KT gives attention to innovation not only in the sense of significant identifiable technological advances, but also from the perspective of the discovery process and its imaginative exploitation (Kirzner 1985). Hence, KT is about the effectiveness of the knowledge value chain, which allows for unrecognized means and ends to be discovered.

Technology transfer (TT) is a related but different subject. TT emphasises information and efficiency rather than knowledge and effectiveness. A TT program is a search/response mechanism that uses technical concepts to transfer technical information and data from the results of scientific research. If implemented with efficiency and speed, an information- and data-oriented approach helps develop applications that solve practical problems in products and processes of an individuated industry.

KT can push the economic performance of a country, region, or industry insofar as it provides a competent guide to the innovation process, which relies on a complex web of relationships. A broad range of competencies to identify, capture, industrialize, and commercialize free flowing knowledge and technologies combine to make KT conducive to economic growth. The higher the quality of these competencies, the more likely an entire economy is to benefit from new venture creations that are superior market entrants, as well as from the successful reorganisation of existing firms.

KT relies upon the action and flow by which largely tacit knowledge, not technology per se, is transmitted among people: from one unit (the source – a single person, group, or organization) to another (the recipient), with all kinds of feedback loops. The process is complex and nonlinear, with a large number of interactions, and is not simply a matter of knowledge passing down a production line that links researchers upstream with their business counterparts downstream.

KT is concerned with the subsequent absorption through which the recipient is affected by the source's experience. How to transfer knowledge that exists in a given unit into another is more than a communication problem that information technology (IT) tools can fully resolve.

First and foremost, KT is an evolutionary process of communicative interaction through "conscious conversation," which gives birth to markets. In fact, markets are conversation. Being part of a culture of conversing in low context communities of knowledge practice involves each participant having his or her say on equal terms, in a nonconfrontational, nonstatus, and friendly and open way. All members set their own agendas based on their passion, and they organise the conversations themselves. New knowledge is created from questions that arise during these conversations, a process that invariably leads to surprising learning and outcomes (Stewart 2001).

Conscious conversation involves human action to construct and transform a mental content, and then human interaction ("action of social relating") for an effective sharing of knowledge, ideas, and experiences. Here, it is possible to conceive of supportive mental tools akin to knowledge resource maps that "show who has what knowledge and what sources are used" (Stanford 2005).

Since human interaction happens in a community and needs behavioral rules that allow relationships to take place without the rigidity of formal contracts, the amount of social capital available in a given community (see the following chapter) is critical to a viable KT process. A vibrant social capital infrastructure founded on the intelligence and interactions of people with shared visions and common purpose invigorates the KT performance.

Exhibit 29: Conscious Conversation

"Conversation comes from the Latin cum versare: to turn or to dance together... Products, inventions and services, great and small, are created through talking. The structure of the DNA molecule was talked into existence by Watson and Crick, in the lab, pub, train and on the river. Linus Pauling, who had done extensive work on the same project, had nobody to converse with (reportedly by choice) and missed out on this major discovery" (Stewart 2001).

Conscious conversation is "a transformational change technique that incorporates deep dialogue skills of reflecting, deep listening, interacting and connecting. It intends to foster common sense, build trust and understanding, and create positive and harmonious relationships among community members. It is familiar to communities of practice in Asia" (Yin and Lin 2002).

The best preceptor is a participative engagement in the conversation between knowledge seekers and knowledge users. Conversations are the sense-making conduits through which knowledge flows (Kilpi 2005). This is where learning dynamics and learning value can be optimised. From this perspective, an effective KT process directs its attention to a "conscious conversation" as the central activity that involves the deployment of a wide range of soft skills.

Exhibit 30: Soft Skills

Soft skills are behavioral and social components of "emotional intelligence" as opposed to "technical intelligence" (Leonard 1998). They encompass values, motivations, attitudes, and emotions:

- Skill in self-awareness (recognizing your own strengths and weaknesses)
- Skill in self-regulation (keeping emotions under control)
- Skill in motivation (having optimism and personal drive)
- Skill in reading emotions and motivation of other people (empathy)
- Ability to build and manage relationships (negotiation skills)

Soft (emotional) and hard (technical) skills plus traditions make up the competencies that permit individuals with different experiences and/or educational backgrounds to intertwine interpersonal relationships.

Exhibit 31: Knowledge Resource Map

"A knowledge map differs from an information or data map by its function or purpose. Generally the purpose of an information map is to show us what we have and where to find it.

"A knowledge map is intended to help us

- Learn
- Build
- Elicit
- Share
- Create
- Regenerate knowledge

"It is one of the tools used to make implicit knowledge explicit so it can be made implicit again.

"A knowledge map includes a text connected to symbols, directions, routes, and other key map elements. The text should be concepts, questions, or ideas – not paragraphs of information or simply objects.

- A knowledge map shows relationships between or among the concepts.
- The value of the true knowledge map can be broken as follows:
- What you have that you need – so you can leverage it.
- What you have that you do not need – so you can eliminate the "fat"' and concentrate on more important elements (the 80/20 rule).
- "What you do not have that you need – so you can obtain it" (Stanford 2005).

Automated tools for developing MindMaps such as "MindManager" allow quick and easy assembly of mind maps and efficient export of such mind maps into commonly used formats.

Knowledge Transfer: Interaction Between Academia and Industry

Interactions between academic research and industry are a cornerstone of KT. From an economic performance perspective, it is vital that knowledge flows from academia into business and society at large.

In the academic context, KT covers the processes of transferring knowledge, research, skills, experience, and ideas within universities, and from universities, to the greater community of users (the business sector and the wider community) to increase economic returns and achieve cultural, educational, and social benefits for society (HMSO 2003: page 39).

This definition embraces both the form of KT and that of TT. These two forms sit side by side under the organizational umbrella of a multidisciplinary entity (known as the "Office of Technology Transfer," which often includes scientists, engineers, economists, and marketers) dedicated to identifying research results of potential commercial interest, and to developing strategies for how to exploit them.

There is a subtle point that descends from how academic research is to be exploited. If industry is considered, a customer seeks specific deliverables, then exploitation implies applied research contracts with industry. In turn, this means that the transfer process is a one-time handout process from the resaerch originator and supplier (research centres and their laboratories) to the developer (industrial client). Eventually, a "lubricant" in the form of a broker or interface might speed up the transfer process (Mitra and Formica 1997: 22). Conversely, if industry is considered a "patron" demanding excellence, motivations to research transfer become more important than modalities. The aim is to conduct research in areas of interest to business (i.e., "applicable" research).

Exhibit 32: Types of Research and Transfer of Research

Pure research is discovery; basic research is the understanding of the mechanism of discovery; applied research is the investigation of the economic utilisation of discovery; and development is the exploitation of discovery.

Transfer of research means transfer of new technologies (technology transfer), transfer of information to markets (information transfer), and transfer of younger specialized personnel (personnel transfer).

Exhibit 33: KT Activities from an Academic Perspective

- Exchange of knowledge through teaching, training, research, or industrial partnerships involving faculty members and students.
- Application of knowledge to social and political issues of the day through participation in advisory boards, government consultations, advice to interest groups, public commentary, and other forms of community service.
- Codification of knowledge through written articles, conference presentations, or patent applications.
- Commercialization of knowledge through the development, exploitation, and marketing of products for the domestic and international marketplace.

Source: Trends in Higher Education: page 78, 2002; Natural Environment Research Council (NERC) UK – http://www.nerc.ac.uk/using/ktcall.shtml

Exhibit 34: Key Principles of Motivation to the Transfer of Research

- The lubrication of the transfer channel cannot by itself accelerate the process in the absence of driving forces. The process is driven by demand rather than supply.
- The demand of innovative products and services is the driving force for entrepreneurs to invest in research: a general criterion should be to pull the innovation rather than to push it.
- To accelerate the transfer, it is necessary for receivers to be strongly motivated and in close contact with their markets. They must also be capable of assimilating the research they receive. It is here that a good educational and training system conveys a high-priority.
- The cost of introducing a new technology is far greater than its transfer price, and is connected to the changes of the firm's organisational structures and strategic assets that usually come together with the adoption of a new technology.

Source: Adapted from Nicoló (1996)

Impediments to Knowledge Transfer

Collected experience shows that transfer processes are subject to a variety of barriers. They include the following:

- An inability to bring together the right competencies.
- An inability to detect those competencies that are highly intuitive rather than consciously perceived.
- Internal conflicts stemming from "professional territoriality" in a given area of expertise.
- Generational gaps.
- Inappropriate identification of the key holders of specific knowledge or content.
- Problems with sharing beliefs, assumptions, heuristics, and cultural norms.
- Deficient or ill-defined motivations and incentives for significantly greater interaction among the parties involved in the transfer process.
- Inadequate mentoring or guided experience.

A great deal of research is needed to examine the barriers that stand in the way of what otherwise would be a satisfactory transfer process.

Routes to Knowledge Transfer

KT can occur via various routes. Processes of integration, collaboration, communication, and commercialization of knowledge are associated either with the softer side of the transfer process, such as sponsored students and contract and collaborative research, or with the harder side of it, such as intellectual property, licensing, and spin-off companies (HMSO 2003: 39).

Knowledge Integration Process

The rationale that sustains this process is that we are no longer in the age of information. Economies are shifting from information to knowledge integration. Hence, the view that the economy is poised to bounce forward has to be built on the economy's knowledge bases. This requires an integrated approach to respond to the new economic and social needs.

University–industry interactions reveal two basic types of relationship that preside over the knowledge integration process:

- *Type A: Transfer of inputs ("supply push")* – A type of relationship that concerns contract research, consultancy, and other university outreach initiatives to business, such as transfer of research, skills, management strategies, and knowledge capital in general.
- This relationship emphasises the supply of input (of a "knowledge package"), lending relatively little weight to the interaction with end users. The crucial consequence of a linear approach is that it neglects the organisational and behavioral characteristics of SMEs, as well as their capacity to absorb the input transferred.
- *Type B: Knowledge transfer designed in a demand-led way ("demand pull")* – This is a coupling type of relationship that has two properties. One property makes the relationship dependent on the needs of business and, therefore, its primary objective is to fit the cognitive characteristics of the recipient actors (Garavelli, Gorgoglione, and Albino: Part 1). A second property is that the relationship is driven by the interplay between the supplier and the receiver of knowledge. The better the interchange, the higher the value of KT, and the more intense the iterative process, which by trial and error produces new knowledge at every stage.

It has been found that knowledge transfer, designed in a demand-led way and capable of fostering ties with knowledge providers outside the region, is to a significant degree important in the following:

- Those regions or countries with a low density of knowledge services (reflecting an inadequate knowledge base)
- Where local firms are learners whose very limited capacity of absorbing new knowledge fields requires a language of communication that reflects the learners'

preferred ways of being instructed about that new knowledge (Tödtling and Trippl 2004; Powell 1987).

In these problematic regions, a critical role could be played by a relationship promoter who would be responsible for gathering subject-specific knowledge that backs local firms and thereby improves the effectiveness of the knowledge exchange (Gissing 2005). Relationship management would help to raise the meager demand for the knowledge and skills available at university sites, redirecting a share of university research to SME needs.

Knowledge Collaboration

Collaboration, along with social cohesion and connectivity, is crucial for knowledge sharing and value creation. The value of leveraging knowledge between partners creates greater wealth and sustainability for us all (see the "Third Law of Knowledge Dynamics").

Knowledge collaboration describes an open process of value creation in which contributing members make every effort to capture all the relevant pieces of knowledge across functions, businesses, and even across nations.

Different tools are used to create meaningful venues for collaboration. The tools described below have been molded over many years of collaborative experience between academia and business. They show two facets: one is a controlled situation (closer to the concept of a contrived consultation) in which each party involved solicits a demand or response from the other component(s); the other is an unstructured, unpredictable, and spontaneous interaction, which promotes cross-fertilization of ideas for prosperous innovation.

Traineeships/Internships

In this organizational form, knowledge transfer occurs by means of interaction between the knowledge provider ("teacher") and the recipient individual ("learner"). The training process enables the learner to use, in a clearly defined context, the knowledge transferred by the source. The provider knows a-priori the solution to a specific problem that the recipient has to solve (Garavelli, Gorgoglione, and Albino).

Knowledge practice includes both project-based placements of students in a company[6] and company employees in an academic lab for the realization of a specific project, which is the mission of a partnership between the university and that company.

Continuing Professional Development

Continuing professional development (CPD) is an important form of knowledge transfer that an increasing number of universities are providing to business employees.

Through CDP, "Businesses can raise the skill levels of their workforce and learn about the latest academic ideas, while universities gain access to the latest developments in professional practice" (HMSO 2003: 122).

Collaborative Research

The collaborative research form of knowledge transfer aims at promoting a context in which academic researchers work alongside company employees to create, develop, and test a prototype based on their reciprocal ideas and which could form the platform for the development of a new product or service, possibly leading to the creation of a new focused on application fields far from the original use of the knowledge transferred.

The company can provide the partnership with staff, the data and equipment necessary for the successful testing and development of the prototype, and with funding. The academic partner can tilt the university action in the direction of offering access to both in-house expertise and its international network of scientists and researchers.

Collaborative research can be carried out in a "collaboratory" – an appropriate lab-type infrastructure that links teams of people from university and companies with disparate cultures, cognitive systems, and skills. In a collaboratory, research focused on specific company problems and scientific research is carried out through the interactions between academic trained corporate researchers and university researchers willing their scientific results to put to practical use.

By providing access to this infrastructure to groups of talented students or postgraduate students who can benefit from the knowledge exchange among the participants, spending more time working alongside academic researchers and company employees on shared problems and projects, new business formation become more likely to occur through spin-offs and start-ups that lead to new knowledge-based enterprises founded by students and graduates, and supported by "in-collaboratory" companies.[7]

Knowledge Communication

Tacit knowledge is not transferable without communication between individuals. To share knowledge, trust and understanding are important factors. Each participant in the transfer process needs to develop autonomous critical capabilities and practices for the purpose of making effective use of the knowledge transferred.

The extent to which knowledge communication is built on the principle of participation (by being evocative and not only informative) indicates how powerful it could be in shifting the current emphasis on information in favor of imaginative ideas to be converted into sound commercial ventures.

At present, most universities are still organized to inform faculty and students about the process of commercial development from academic research. Workshops and seminars help to communicate an understanding of this process, but their informative content is too limited in its scope: it does not address the recipient's need to acquire that autonomous practice, which would allow it to play in the realm of imagination where the information is interpreted and turned into knowledge in action.

The much-vaunted university channel of knowledge communication is at the intersection between disciplines, both technical and business, and capable of melding the worlds of science and industry. Funding interdisciplinary chairs that focus on both technical and business topics is a first step toward that and would give fresh weight to the question of how universities can contribute to effective knowledge communication.

Knowledge Commercialisation

An increasing number of academic institutions are aiming to convert knowledge creation into economic knowledge that can constitute a business opportunity.

In this respect, some universities have set up one-stop centres to guide faculty inventions and scientific research through the commercialization process. These centers are focused on the following:

How to assess the commercial applications of the results of a research project?
How to effectively formalize them into a business plan?
How to identify the best way (product, service, technology) to commercialize the results of a research project?

UK universities, for instance, have established science enterprise centers "to foster the commercialization of research and new ideas; to stimulate scientific entrepreneurialism; to incorporate the teaching of enterprise into the science and engineering curricula; and to act as centres of excellence for the transfer and exploitation of scientific knowledge and expertise" (European Commission 2004).

Incubation of Research-Based Start-Ups

Knowledge transfer involves new business launches or identification of new business opportunities within existing organizations.

Universities and other higher education institutions that put in motion processes of knowledge transfer are often also interested in embarking upon a process of incubation ventures through which knowledge-based opportunities flow across conventional intellectual and business borders. By doing so, they support ventures that originate from scientific research.

The incubation process generally takes place in a physical and organizational infrastructure called an "incubator," which measures the success of higher education not only in graduates but also in faculty-student promoted real business start-ups. Scientists, academic researchers, and talented students who perceive practical implications from their findings often lack the strategic vision and profit-seeking approach required of a would-be entrepreneur. The incubation process brings together, in a single organisation, these entrepreneurial scientists, researchers, and students, and enhances their ability to interface knowledge and innovation. Research findings and novel technologies, which are the result of their curiosity-driven research projects, are redirected toward business concepts that can be converted into viable commercial products and services.[8]

Developing spin-off firms based on sharing university potential is not the sole role of the incubation process. The same process can also spin in creative ideas from local businesses and help to form partnerships for new venture creation with the pool of knowledge-rich scientific and technical personnel, and talented students, backed by the incubator infrastructure and its support staff (Powell et al. 2000: 11).

A good number of university spin-offs that have the status of a joint closed stock partially or fully owned by both an academic institute, which is committed to the exploitation of its research results, and one or more scientific entrepreneurs (entrepreneurial scientists included) may not prove to be sustainable. Rather, this increases the likelihood that something negative will occur, and therefore the propensity of universities to shift the emphasis from developing commercially viable academic spin-offs to being much more focused on licensing.

MIT, a leading institution in the transfer process, has been a pioneer of policy efforts designed to tackle the issue of licensing. As observed by the Lambert Review (HMSO 2003: 67), "Unlike many UK universities, MIT has no business incubation activities at all. The strategy of the technology licensing office (TLO) is to encourage as many invention disclosures as possible from faculty members by minimising the barriers to disclosure – currently MIT discloses about 450 inventions per year. MIT's TLO then licenses these inventions as nonexclusive or exclusive licences to industry and local venture capital firms. Rather than getting involved in the complexities of spinout formation, the TLO provides a shop window for industry to view its IP and agrees as many licence deals as possible."

A licensing policy opens up opportunities for incentives that motivate inventor academics to seek patents as a means of maintaining control over future research (Strandburg 2005).

Exhibit 35: Incubation Process

The process from creative idea generation to full commercialization is articulated in two consecutive phases. First, there is the preincubation phase bridging the gap between the disparate cultures that must be integrated in view of gearing perceived opportunities more closely to the marketplace. Through action learning and team-based pre-start-ups, preincubation leads the way in ensuring that talented would-be entrepreneurs learn how to change their ideas and projects into business processes for developing a new product, exploiting existing know-how, and accessing others' technology or skills. The emphasis is on acquisition of creative and practical skills by motivated individuals who could create their own companies from the business activities they have pursued in the learning network.

Preincubation is followed by the incubation phase, rooted in a business milieu where large endowments of intellectual assets and a wide range of specialized physical infrastructures are available.

Notes

[1] See, for instance, how Andrew Wiles has proved Fermat's last theorem, as told by Singh (1997).
[2] "Information entropy measures the content of a message through the news or surprise it contains – the number of unexpected bits" (Gilder 2002).
[3] Gadiesh and Gilbert (1998) refer to Microsoft's Windows and Intel's microprocessors as examples of choke points in the computing industry.
[4] See "The Learning Age: A renaissance for a new Britain", Green Paper by the UK Secretary of Statefor Education and Employment, London 1998, and "Connected Intelligence Workshop" directed by Derrick de Kerckhove, Mediartech 1999, Florence, January 24–28, 1999.
[5] See, for instance, how Andrew Wiles proved Fermat's last theorem, as described by Singh (1997).
[6] See, for example, the Shell Technology Enterprise Programme (STEP)_a nationwide scheme which provides placements for undergraduates, mostly during their summer vacations, to work on a project that meets a specific business need in an SME. All students receive a skills assessment package and three days' training from their local provider (usually a business support agency or university) to enable them to record the transferable skills learnt during their placement. Many businesses receive contributions towards toward the cost of the placement from local business support agencies, which play an active role in helping them to define and quality-assure the project (HMSO 2003: (HMSO 2003: 120).
[7] See the program to establish Co-operative Research Centres Centers in Hungary. One of the objectives of the Hungarian R&D and innovation policy is to promote R&D in enterprises and their collaboration with universities. This aims to promote joint R&D actions undertaken by universities and enterprises and appropriate transfers, which may lead to new processes or products. Objectives: Objectives: To create, or to strengthen the operation of, research centres centers allowing the formation of integral ties between Hungarian college and university (higher) education institutions, other non-profit nonprofit research institutions and the enterprise-business innovation sector.

This, in turn, enables the strategic integration of education, research and development, knowledge and technological transfer can be realised. As Co-operative Research Centre (CRC) recipients, Hungarian universities and colleges can submit bids, individually, jointly or in a consortium with enterprises. The leading institution of the consortium may only be an establishment accredited by the Hungarian Accreditation Committee for Ph.D. training. CRC proposals should be submitted exclusively with the participation of business partners. The centre to be established can be an independent legal entity or a separately financed, economically independent unit – within the organisation of an institution of university or college education.
The proposal should detail a strategy for long-term (minimum 3, but preferably 6–9 years) research, training, plus knowledge and technological transfer, developed jointly by the participating partners and supplemented by the business plan required for the operation of the centre. Source:http://europa.eu.int/comm/enterprise/enterprise_policy/enlargement/cc-best.

[8] The overriding concern is the conflict of interest that develops as research teams give birth to spin-off phenomena. As Strandburg (2005: 64) has observed, "Commercialisation of spin-offs of curiosity-driven university research may involve the active participation of the scientist inventor. It is not clear what impact the involvement of scientists in such entrepreneurship is likely to have on the market for curiosity-driven research. One salient concern is that an entrepreneurscientist might seek to suppress the work of another scientist if that work had the potential to threaten the commercial success of his entrepreneurial project. The basic research community might effectively avoid this potential distortion of the curiositydriven demand function by using more stringent conflict of interest screening of peer reviewers. Scientists with commercial stakes in enterprises related to particular areas of curiosity-driven research could be precluded from reviewing proposals and publications in those areas."

Chapter 4
Industry and Knowledge Clusters

Trust is the bandwidth of communication.
(Karl Eric Sveiby)

Clusters can be likened to the living organisms in the rainforest populated by a vast number of species with wide range, such as the parasite, the predator, the symbiont, and the table companion, which will be described in the following paragraph. In the rainforest internal design, which means relationships, is what is valuable. The quality of relationships in a cluster is heavily affected by social capital, which tends to reflect the special characteristics of a particular cluster community in terms of mores, manners, and in the hearts and minds of those who live in it. Social capital thus differs from cluster to cluster, and each cluster is unique – as is the case in a rainforest, where only one of two identical organisms survives. Clusters are "one-of-a-kind," and not "one-size-fits-all," living organisms.

In 1890 Alfred Marshall published his *Principles of Economics* in which he included a chapter on "Industrial Districts," as he defined the geographic concentrations of specialized industries. After the Second World War the term "industrial district" has been applied to the emergence of economic specialization in specific communities in Italy, mostly in the northeast regions of the country. A century after Marshall's book, Michael Porter's neo-Marshallian cluster concept burst on the scene through a series of seminal articles (Porter 1998a, b, 2000). By the time this book come to fruition we were at the height of a cluster boom. Economists, sociologists, demographers, ecologists, biologists, and policymakers debate cluster theories and practices. Every year, hundreds of weighty tomes and books on clusters pour out of universities and think tanks, assessing the significance of past events, investigate the consequences and discuss what will be done in the future. International conferences are held on the topic, and specialized, government-driven agencies have been established with the purpose of creating clusters and improving cluster development. All in all, it is fashionable to say "we are all cluster supporters now."

There are also institutions such as the European Commission's Information Society and Media Directorate General that see the cluster phenomenon from the perspective of the early literature on growth poles, holding the view that there is a certain kinship between clusters and growth poles [as per the seminal research on

T. Andersson et al., *Knowledge-Driven Entrepreneurship*, Innovation, Technology, and Knowledge Management 102, DOI 10.1007/978-1-4419-1188-9_4,

growth poles going back to Francois Perroux's work in the 1950s (Perroux 1950, 1955)]. *A Newsletter on New Methods of Work & e-Commerce* (July 2002, No. 17) published by that Directorate General stated that the growth pole "was concerned with the transport of physical goods and, for this reason, stressed the importance of physical proximity to the eventual success of a given growth pole. Work on clusters also stresses the importance of physical proximity, but with a much greater stress on face-to-face knowledge flows than on the flow of physical goods. Here, the paradigm case of a successful cluster is Silicon Valley. Both the flow of goods and the flow of knowledge, however, share a common requirement to mitigate the negative effects of spatial distance… we integrate the growth-pole and cluster concepts into a single term: the growth-node."

Seductive though the cluster concept is as an engine of growth, much about it is problematic. The terminology is used in quite a cavalier manner, while the nature of the concept poses a problem of definitions. Indeed, the cluster concept in social science belongs to speculation rather than science. Scholars and experts in this field have looked increasingly at the "softer" attributes of clusters – trust, social capital, leadership, teamwork, networking, creativity, and the like. As they have done so, it has become increasingly difficult to assess with any rigour whether their hypotheses are true or false. As shown in the Cluster Policies Whitebook (Andersson et al. 2004), clusters can serve to strengthen developmental and defensive forces and need appropriate policy frameworks and stakeholder relationships to evolve in a favorable manner. Recent attempts by the OECD to construct a trust indicator show that the research community has taken a course of action that is in harmony with the needs of a scientific approach to the cluster economy.

Exhibit 36: Trust and Social Capital

Trust is a critical component of the collaborative formula for those who should build bridges in a field as subtle and ambiguous as that of know-how, know-what, know-why, know-whom, and know-when. When compared to trust-promoting groups, vested interests seem to be having little success as knowledge intermediaries.

Trust emanates from a personal feeling. It is a distinguishing characteristic of a relationship. Trust is a social dividend that springs from the individual need to preserve his/her own reputation in the community (Kay 1995). Thus "trust means you aren't going to find a way to take advantage of me. You are not selfish for your own self. The partnership comes first" (Uzzi 1997).

Social capital is a key factor in cluster formation and development. An international expert survey has identified social capital as being among the top attributes of a successful cluster (The Clustering Alliance Newsletter, July 2002, No. 31).

Social capital is embodied in the relations among people for collective actions with mutual benefit. It refers to a learning process of sharing, through everybody's interaction, a set of conventions, norms, values, beliefs, aspirations, common working styles, and overlapping objectives (Coleman 1990; Stanford 2002).

(continued)

Exhibit 36 (continued)

The OECD definition of social capital "includes such structural and psychological elements as the networks of personal relationships and sense of mutual understanding that enable people to live and work together effectively."

"Social capital is associated with greater trust, co-operation, reciprocal engagement and social cohesion. Furthermore, social capital can enhance the rapid diffusion of knowledge between individuals, communities as well as within and between firms" (OECD 2002; OECD/Government of Canada 2003).

The PRISM REPORT 2003 by the European Commission Information Society Technologies Programme defines social capital "as the set of collective (in the sense of shared) intangible assets available in a territory (a city, a region, a country, a set of countries). Collective intangible assets allow communication and exchange to take place without rigid, formal contracts because they provide behavioural rules (formal or informal) that avoid free-rider problems or other abuses of loose contractual relationships" (Eustace 2003).

Social capital breaks down into traditional and spontaneous sociability. Traditional sociability "can be said to be loyalty to long-established social groups, [as in the case of] the medieval producers," such as the Hanseatic League guilds. Conversely, spontaneous sociability "is the ability to come working together and cohere in new groups for common purposes, and to thrive in innovative organisational settings" (Fukuyama 1995).

Spontaneous sociability is associated with strong trusting relationships that extend well beyond family ties. Trust reduces the costs of monitoring contracts and allows risks and information to be shared. People disseminate knowledge and valuable information between firms as they used to do within them. When the success of the partnership is at the top of the agenda, no one will seek ways to take advantage of the others and will therefore put self-interest aside, be it his own interest or that of his family (Uzzi 1997).

The type of sociability that will prevail in a given community depends on the values and beliefs, unwritten norms and conventions that the individuals in that community share in their everyday interactions. For traditional sociability, the common trait would be kinship and family ties where people trust only their close relatives. In communities that live in a social climate of familist behavior, individuals mistrust one another. Here, the entrepreneurial energy generated by the founding fathers tends to be dissipated and degenerates into a syndrome of the "third generation" once the baton passes from the founders to their descendants.

High-trust communities are those in which nonkinship relationships prevail and do not rot in interest groups that support restrictive practices. They accomplish a "chain reaction" of social capital between (1) the individuals' spontaneous ability to associate with each other, (2) the degree to which the communities share norms and values placing common interests before individual ones, and (3) trust that comes out of such shared values and, in turn, feeds spontaneous sociability.

The accumulation of social capital underpins the simple proposition that the path to prosperity could be trodden by almost everybody. In fact, shared prosperity

(continued)

Exhibit 36 (continued)

is the striking feature of a collaborative economy, while widening inequality poses a threat to society.

Trust cannot be catered for; it represents a subjective state of affairs. In digital communication, the traditional means for building trust are breaking down, while there is also a failure to put in place the most basic means for securing user authentication. Along with a lack of means to trace and validate data, this has resulted in a breakdown in security and trust in digital communication. To address the situation, today's data-centric approach needs to be replaced by a user-centric model, which starts out with the individual user and person and which can help to enable trust by allowing the indvidual user to gain control over his or her identity and over who has access to associated communication (Andersson et al. 2006).

The argument that people and their interpersonal relationships are the lifeblood of clusters is an entrenched contention, and even a cherished belief, in the literature on clusters. But careful scrutiny is required for avoiding such confidence being misplaced. By focusing our attention on the forces that create and develop clusters, a line can be drawn between relations that portend a loss of personal autonomy and others that, conversely, magnify the independence of each and every individual inside a given relation.

In the first instance the individuals are "insiders" subordinated to the norms and procedures of the organizations to which they belong. They act "in the name or on the behalf of" the organizations and end up doing what the organization has resolved to do. "Industry clusters" are what we call those clusters where it is not the individuals but the organizations that hold everything together.

In the second instance, the clustering process is rooted in lateral or longitudinal relations, which refer to interpersonal exchanges developed by individuals in their own right (to wit: "outsiders") rather than as representatives of organizations. These individuals form affinity groups, which evolve in what we call "knowledge clusters."

Exhibit 37: Affinity Groups

Theodore Zeldin in his "Intimate History of Humanity" noted that affinity groups are those intermediaries or catalytic agents that "can create new situations and transform people's lives by bringing them together" (Zeldin 1994: 155). By the way, it can be noted that "affinity" was the word used by Newton and Goethe ("Elective Affinities" is one of Goethe's most famous books) in relation to the chemical phenomenon (later known as catalysis) of two substances that combine to form a third one.

Affinity groups give birth to "intermediate-sized audiences and communities focused on common interests" (Kelly 1998), which exploit the vastness of the borderless Internet for selling into niche markets on a global scale (Fig. 6).

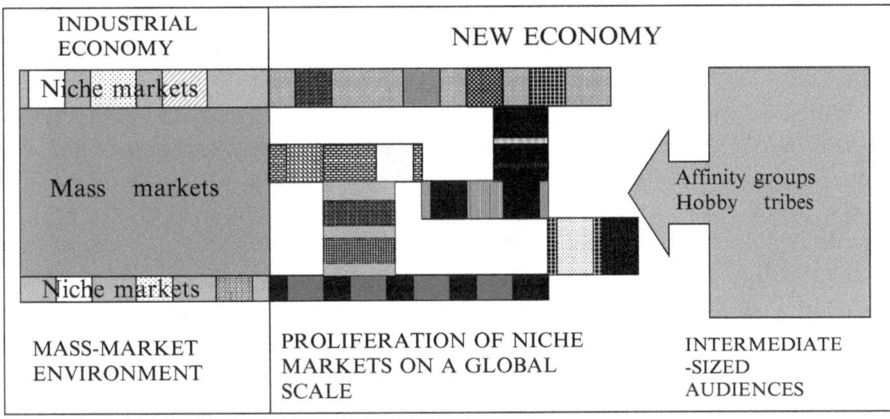

Fig. 6 Proliferation of niche markets on a global scale [Source: Adapted from Kelly (1998)] Internet technologies encourage the growth of markets that could not have existed in the industrial economy. Greater numbers of small companies are able to find niches where niches were not possible before.

The global scale provided by the Internet is not equivalent to a mass market. Of course, what entrepreneur does not dream of reaching the maximum number of customers? But going for big numbers implies deep pockets and strong intangibles (brands, reputation, and services) and the capacity to compete with many rivals that are fighting to attract the attention of the same customers. For this reason, newcomers and small businesses are at odds in a mass-market environment.

Those who invent or detect affinity groups and are thus prompted to address niche markets will likely earn a profit at an earlier stage. In contrast, others fascinated by the big numbers will generate losses beyond the amount foreseeable in terms of high start-up costs. In turn, this means that they will have to bear the burden of a negative gross margin on variable costs. A generous profit margin might be attainable over time, but the appeal of this carrot, combined with low entry costs for Internet-based businesses, would attract new entrants and thereby erode that advantage.

Public authorities have shown themselves eager to use their powers to intervene in the organization of the cluster economy. Thus, cluster policymaking has come to be accepted as an established fact and is now entirely within policymakers' scope. Public authorities subscribe to the principle that individuals' spontaneous and uncontrolled efforts provoke market failures or markets that produce only short-term results. Cluster building is a long-term process that requires a strategic enabling factor. This is associated with a mode of governance based on policymakers' deliberate actions.

In this respect, our view, in line with the sort of criticism that Martin and Sunley (2002) have voiced, is that the principle of government intervention for cluster-building reflects the influence of Porter, his disciples, and followers, who have first created an intellectual product called a "cluster" and then have turned it into a successful brand, which has been extensively sold to policymakers and a vast group of associated practitioners and consultants.

The shape of public policy has, unfortunately, taken something of a back seat in the cluster research agenda. Following on from the pioneering work of Vernon Smith (2000), the 2002 Nobel Laureate in Economics, in the field of experimental economics, efforts should be made toward the creation of laboratory experiments to test different institutional environments for clusters. Many questions remain to be answered. For example, is there really a need for cluster policymaking? What types of social norms, rules, and choices would best facilitate the pace of development through a cluster-building process? And what if participants in the economic arena are willing to use market mechanisms to find their own way to change themselves and their communities?

Industry Clusters

Businesses have clustered into networks of various sorts throughout history. The medieval guild system was a primitive networking exercise.

US National Commission on Entrepreneurship, USA

Some local communities are in better shape nowadays – economically and socially – than for generations. This unprecedented economic development has not been inspired by governments, but is the result of organic growth.

The clustering process is fed and catalyzed by community-rooted entrepreneurial heroes and probusiness academics too. They are the "food molecules" without which the "cluster reaction," which gives shape to a collectively autocatalytic system, would only take place with great difficulty. Willingness and capability of the key players to generate new relations in the system determine the pace of growth of the cluster.

An industry cluster – as a self-sustaining, self-reinforcing group of businesses has been termed – generally centers around a core of highly specialized enterprises in the same industry enriched by close links with supporting firms that produce what the former need. Core businesses are privately owned and are often family-run businesses and family-controlled groups, few of which are listed on the stock market. Industry clusters are governed by the rule of the authority and delegation under the jurisdiction of the industrial leader(s). The idea is that companies should concentrate on a few processes while buying in the rest from third-party suppliers under long-term service agreements. This is the fact that features most prominently in the industry cluster economy.

Exhibit 38: Autocatalytic System

A collectively autocatalytic system is defined as "one in which the molecules speed up the very reactions by which they themselves are formed: A makes B; B makes C; C makes again. Given a supply of food molecules, the network will be able to constantly re-create itself" (Kauffman 1995).

Figure 7 highlights the formation of the autocatalytic system of the packaging machinery cluster in Bologna, Italy (see Appendix 6).

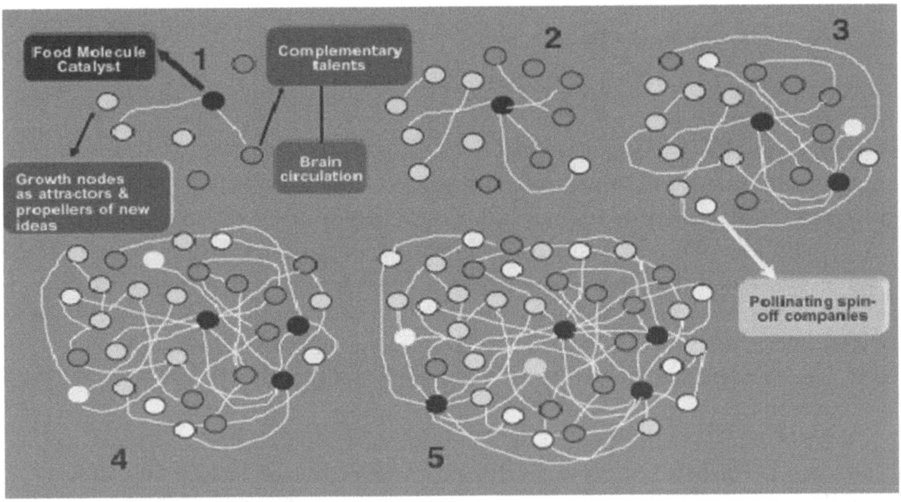

Fig. 7 Formation of an autocatalytic system: the case of Bologna packaging machinery cluster Dots "buttons" (growth nodes) are at random connected by an increasing number of lines *"threads"* (edges). Buttons symbolize the key players: artisans, technicians, companies, and "food" molecules. Threads symbolize relations between players, who send out new threads at each node. "When there are very few threads compared with the number of buttons, most buttons will be unconnected. For large numbers of buttons, as the buttons pass a threshold of 0.5 a phase transition occurs; most points become connected in one giant component, and a giant cluster suddenly forms. As the ratio passes 1.0, closed pathways of all lengths begin to emerge" (Kauffman 1995). The Bologna packaging industry has overcome the transition phase and is now like a giant component, whose "rate of growth slows as the number of remaining isolated buttons and isolated small components decreases" (Kauffman 1995). This is why the builders of automatic machines are compelled to innovate in the relationships with their customers in such a way as to accelerate the growth rate by selling a complete package of services, the machine being just one part of the package

Exhibit 39: Industry Clusters

Industry clusters can be circumscribed using any combination of a broad spectrum of criteria.

An industrial cluster can take different but also overlapping forms. Both faculty researchers and practitioners have provided several definitions. For instance, close geographical proximity of interconnected industries such as businesses that share similar workforce, input, or infrastructural needs gives rise to a regional cluster.

Industrial districts are local clusters where spatial concentrations of micro-to-small inward-looking local firms and small-to-medium companies predominate. The latter often hold an international niche leadership operating close to one

(continued)

Exhibit 39 (continued)

another and producing similar goods. These firms are also engaged in interdependent production processes. This type is strongly represented in Italy, mainly in the northeast regions of the country, where industrial districts are a striking example of an evolutionary environment from the original structure of all-equal-small firms to a set of "wedges" with a prime company at the top.

The OECD-DATAR first World Congress on Local Clusters pointed out similarities and differences characterizing local clusters. The former encompass a high degree of market sensitivity, flexible means of production, and proximity-based interactions between firms. The latter include economic, cultural, and social assets, services, and levels of trust, latent conflicts, and cut-throat competition among firms (OECD-DATAR 2001).

Industry clusters are focused on buyer–supplier relationships, common technologies, and distribution channels. By sharing functions such as order taking and shipping, companies in supply chains are able to reap economies of scale, thereby lowering chain members' costs. Inside the chain, small and micro firms specialize in only one phase of the cycle – resembling the configuration of a Roman phalanx, a body of soldiers in close formation for fighting. In the case of the Italian districts, collaborative specialists are the primary participants in the phalanx. These are low tech but highly skilled suppliers tightly closed to their primary customers.

Along the learning process of collaboration, some firms assume the role of local decision makers. An evolutionary form of coordination then emerges, configured like a wedge. The leading company at the tip of the wedge embraces problem solvers or comakers involved in the early phases of the leader's development process and a platform of secondary and tertiary subcontractors underpinning the new business configuration. At this stage the architecture of collaboration shows a large gray area between propensity to binding commitments and propensity to opportunistic actions.

Geographical proximity allows intense interaction and enables a large amount of information to be exchanged within and between established webs of complementary or interdependent activities that form tight supplier–customer links through a common vision, leading to formal and informal networking. The core spawns new businesses and arouses plenty of rivalry between companies. Yet cooperation and trust occur at the core. Supporting firms include banks, accountants, lawyers, designers, freight forwarders, component producers, suppliers of specialized inputs and their subcontractors, providers of raw materials, and equipment supply and service companies.

In this type of environment joint projects between specialized organizations (e.g., industry associations and insider lobbies) are common. These generate public–private partnerships that often seek support from government programs.

The prevailing business species in an industry cluster community is symbionts seeking connection with other similar companies. The enlargement of the habitat populated by the symbiotic species facilitates more and stronger complementary links within the cluster. There are other species, such as the parasites who live at the expense of other species, the predators who hunt other species, and the table companions who bring no dowry with them and do not cause turbulence or interact with other species (but they are rare).

A profound technology culture and a widespread entrepreneurial spirit from within the business community are at the foundation of industry clusters' success (Appendix 6 describes the success story of the packaging machinery cluster in Bologna, Italy).

Industry clusters are remarkably resilient. "The speed at which labour, land and capital are reallocated is breathtaking," the experts say. Yet, cluster decline, albeit genteel and relative if not absolute, is a potential problem or, worse, an inherent systematic feature of industry cluster dynamics. In fact, success through organic growth induces lock-in- or path dependence, which is a consequence of positive feedback that confines the cluster economy within the entrenched practices of thinking and doing things (Arthur 1989). The achievements of the past change into losing values. Know-how and skill, experienced workers, specialized infrastructure, interfirm linkages, strong political support and, in general, all the institutional, social, cultural, economic, and technological factors that once made it a successful cluster, then cause lock-in dependence.

For example, Bologna's packaging cluster suffered from lock-in dependence dating back to the Industrial Revolution when the city was at the peak of its success in developing the wool and silk textile industry (for which it was renowned in Europe as Lyon). Precepts and forms of the new scientific and technology domain, together with workforce practices that the Industrial Revolution introduced, caused a deep and prolonged recession of the Bolognese economy, influenced by a "cargo cult science" of people, who had difficulty in understanding the principles of the incoming scientific age.

Nowadays, when a multitude of businesses are confronted with the leap from the industrial to the knowledge economy, from the production of "atoms" in form of machines that perform "cold" or unintelligent functions to that of "bits" associated with machines that even affect our very culture – in other words, from "making things" to "think-oriented, ideas-based businesses" – reliance on past successes makes a cluster vulnerable to lock-in syndrome and "cargo cult." New entrepreneurial heroes are heralded as essential to mastery of a new domain. Good character actors replace the protagonists, but the shape of things to come is traced by the emergence of new leading personalities who are willing to change the fabric of traditional mental habits and conventional ideas supported by people with similar thought processes.

Policymakers and even businesspeople relabel knowledge clusters the existing industry clusters since a new knowledge-intensive industry has been added to them. From this view, people believe that they can buy pieces of knowledge from a knowledge firm as they usually buy goods and services. What is missed is the importance of managing knowledge for doing current businesses in a different way or for introducing innovative activities. In this respect, communities, which are free

Exhibit 40: Lock-in- or Path Dependence

This syndrome has been recognized by several authors. For example, according to Martin and Sunley (2002): 28), "The competitive strategies of firms in clusters, which are initially highly innovative compared to firms outside clusters, tend to converge (for example through mimetic and normative isomorphism) and to be less innovative over time because cluster firms define their field of competition as the cluster to which they belong, rather than as the wider external industry. This restricted collective perspective gives rise to competitive 'blind spots' which limit cluster firms' innovative potential, strategic positioning, and ability to anticipate and react to industry-wide shocks."

Exhibit 41: Cargo Cult Science

"Cargo cult science" is an expression used by the Nobel Laureate Richard P. Feynman, the father of nanotechnology, to describe the behavior of the South Sea Islanders after World War II.

"During the war they saw airplanes land with lots of good materials, and they want the same thing to happen now. So they have arranged to make things like runways, to make a wooden hut for a man to sit in, with two wooden pieces on his head like headphones and bars of bamboo sticking out like antennas – he is the controller – and they wait for the airplanes to land. They are doing everything right. The form is perfect. It looks exactly the way it looked before. But it does not work. No airplanes land. So I call these things Cargo Cult Science, because they follow all the apparent precepts and forms of scientific investigation, but they are missing something essential, because the planes do not land."

Source: Feynman 1999: 208–209

of the cumbersome aftermath of the Industrial Revolution, are in a unique position. Their ability to detect and manage knowledge sources, flows and processes instead of objects gives them instinctive and intuitive ways of understanding the business context of knowledge issues. These types of communities challenge the ways industry clusters behave and think.

Industry Clusters in the Web Age

New electronic markets reshape the cluster system. They modify the rules of the game and cluster viability and sustainability. Companies that evolve an unrivalled ability to monitor every movement on their Web sites will make the most from unrestricted access to connections worldwide. New markets, new services, and new revenue streams will be created from the potential of unbounded-by-geography connections. Small firms can find niches that were not previously possible. Relationships with customers and suppliers will be redefined.

Five-hundred years of market creation have seen the transformation of the entrepreneurial economy from once-a-week village markets to 24/7 global electronic markets (see Appendix 7). Innovative entrepreneurs are making the most of the potential offered by global connections. They are active members of electronic business communities that themselves are "networks of suppliers, distributors, commerce providers and customers that use the Internet and other electronic media as platforms for collaboration and competition" (Tapscott et al. 1998: 19).

The digital revolution reduces transaction costs, making uneconomic the size and vertically integrated organization of the modern industrial corporation. Returning to Ronald Coase's theory of firm, which holds that the main reason why firms exists is to minimize transaction costs,[1] the fact that the Internet reduces these costs means that it also shrinks the optimal size of firms. The rules of what has been named the "law of diminishing firms" (Downes and Mui 1998) will be carving out the form of the new ventures. Within the Internet, the key to market dominance is not the size but a bunch of qualities that have much to do with leadership style and value-based initiatives. Successful firms have the qualities of being lean and swift enough to bring producers and customers closer together and build tighter relationships between them.

The Internet has started to redefine cluster-based relationships. For example:

- Clusters are traditionally like a convoy system, which means that companies in the same sector move forward at the speed of the slowest. Will this still be true in the e-economy? How many businesses fear that the spread of the Internet will provide little respite from the onslaught of competition?
- Clusters encourage the rapid spread of best practices. Can one expect speed to increase in the e-economy? Will best practices by imitation be sustainable? Might their diffusion impede innovation because "best practices are by and large about efficiency, not innovation" (Horibe 2002)?
- How will the e-economy change the role of social capital in producing economic capital? In clusters such as the Italian districts social capital comes spontaneously from face-to-face informal contacts. Will clusters' family companies have to reengineer their social capital formation to embrace and invest in wider networks, virtual and hybrid? Models that enable mirroring of real world conditions and relationships are now increasingly looked for.[2]

From Physical to Conceptual Marketspaces

The Internet and its use challenge the role of space and geography in clusters and also reinforce the value of human interaction between customers and suppliers, buyers and producers, and consumers and sellers in defined spaces. The Internet effect is often the creation of a new area of collaboration in the fields of buying and selling products and services as well as intangible value generation.

An industry cluster is a business community rooted in a specific territory where feel and touch are the layers on which trust is constructed. The trust so generated is the foundation for conducting transactions in markets that are bound by limited, physical dimensions. Entrepreneurs, employees, and other business-people interact with a fixed and often small number of other people in close, geographical proximity.

Today, a new form of business community is emerging to replace (or create a hybrid) a physical marketplace with a virtual, conceptual marketspace (Fig. 8). The latter is an electronically created business environment that is unconstrained by

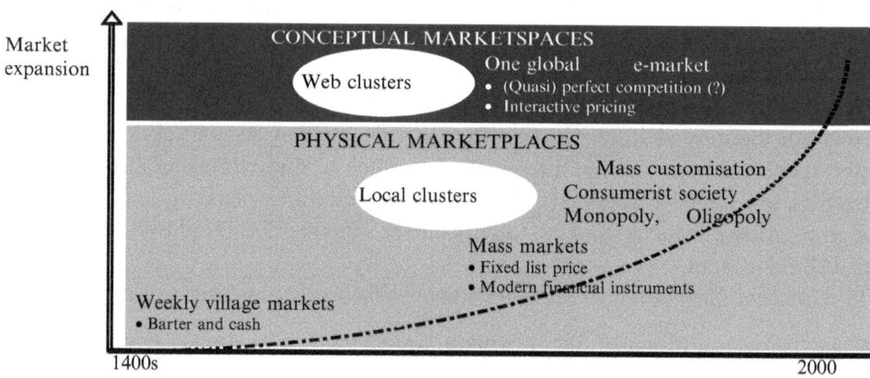

Fig. 8 From physical marketplaces to conceptual marketspaces
Two major clusters of innovations have marked the market's life history over the last 100 years. A device known as the King Road Drag was invented at the turn of the 1900s to make rural roads good enough for a mailman to pass every day. This key advancement in rural roads was the "food molecule" that nourished the formation of a cluster of innovations that gave rise to the mail-order industry. "Now a farmer could look through a catalogue that contained many thousands of products and have them delivered right to his door... You no longer needed to sell to the consumer through actual stores made of bricks and mortar. You could build a virtual store!" (Gladwell 1999). Within few years Sears, Roebuck's Big Store of World Wide Stocks at Economy Prices comes to consumers in the form of a catalogue.

At the end of the twentieth century, the bar code, the Internet, and the last-mile delivery systems merged to form the second cluster. These three innovations have brought radical changes in the ways of warehousing, ordering, and delivering goods. Overall, they have paved the way for a substantial migration from manual, paper-based solutions to electronic purchasing. In particular, the Internet has affected the front end of the operations, knocking down the transaction costs. The bar code has changed the back end into automated, bar-code-driven warehouses. The last-mile delivery systems are an innovative infrastructure made up by planes, aircraft antifog navigational equipment, vehicles, and satellite links that guarantee just-in-time shipping. The bar code and the last-mile delivery systems all together have contributed to cut the average transaction time – that is, the time between the moment of ordering and the moment the goods are received – from 3 weeks to 4 days.

Even though concrete savings can be realized by cutting transaction costs and transaction time, customer-service costs are still there. The click of a mouse cannot erase the need for human interaction.

location. It has been defined as a "a website that allows businesses to buy and sell industrial products and services using a standard web browser. Buyers can post requests for quotes, while suppliers can create online catalogues. Many sites have additional features such as industry-specific news, auctions, and message boards" (Nairn 2000). Participants in the marketspace can weave relationships by means of "experienced" and not only "spatial" proximity, therefore taking advantages of new business opportunities beyond the familiar boundaries.

New forms of communication and new means of distribution might contribute to an increase in price transparency (e.g., online buyers ought to find it easier to discover what sellers are charging) and competition, make price comparisons easier (e.g., sellers ought to find it easier to ascertain what other sellers are charging), and redefine relationships between customers and suppliers. However, there are many factors that conspire to pose a threat to competition, such as those listed here:

- Cyberspace people seem inclined to gather. Thus, groups of buyers and sellers tend to converge toward online exchanges that become dominant. Dominant exchanges could lead to collusion practices, such as price fixing and discrimination against rivals.
- Big industrial and retail buyers aggregate their purchasing requirements. US antitrust officials estimate that a combined purchasing power of more than 20% of a particular market causes damage to competition in terms of creating an oligopsony power that lowers output by pushing prices below competitive levels.
- Participation in an online exchange on exclusive terms (i.e., not to work with other exchanges) stifles competition because competitors are locked out.
- Biased online markets are not alone in their potential to hinder competition. This could also happen through the low-cost and high-effective vehicle of collection and use of information in the cyberspace.

Knowledge Clusters

A desire to gain access to expertise, shared ideas, and learning from one another pushes each individual in the direction of a community of knowledge practice, the atomistic element of knowledge clusters (KCs – see the case of *The Knowledge Management Cluster*), which are molded out of communities of knowledge practice coming together.

If in industry clusters the organizations give orders to people on "where to go," then in knowledge clusters the individuals are able to use their knowledge autonomously and effectively in taking decisions. The distinctive features of collaboration between individuals without the hierarchical ties that bind industry clusters are openness in exchanges (there is a gray area between insiders and outsiders as both use a lingua franca made up from several languages), an ability to question the routines and, as a consequence, the radically innovative mental tools and experiments to turn knowledge into innovation.

Given that interpersonal relations are the distinctive nature of knowledge clusters, the goal of clustering is not to build alliances among firms. It is to build relationships among people. Individuals, not businesses, are therefore the KC players. They include entrepreneurs, aspiring entrepreneurs, scientists, researchers, technologists, investors, professional service providers, and local development officials. Entrepreneurs and aspiring entrepreneurs are the KC's core audience.

More complex configurations include knowledge pools whose members "behave like swarms of bees or flocks of birds aligning with one another and flying in for-

Exhibit 42: Knowledge Clusters

Knowledge clusters are formed by individuals who see themselves as "groups of equals" of knowledge workers whose common purpose, as suggested earlier in these pages, is to change knowledge into innovation and derive business value from it. Those individuals invest in "untraded" or noncommercial factors of interdependency, such as educational attainment, communicating and interpreting knowledge, and attitudes to industry (Storper 1995).

Knowledge clusters, which are endowed with agglomerations of specialized, mutually complementary, and reinforcing knowledge assets in the form of knowledge stocks and knowledge flows, exhibit self-organizing, learning-driven, dynamically adaptive competences and trends in the context of an open systems approach (Carayannis and Alexander, 1999; Carayannis and Sipp, 2005; Carayannis and von Zedtwitz, 2006; Carayannis et al, 2006; Carayannis and Campbell, 2006; Carayannis and Ziemnowicz, 2006; Carayannis and Formica, 2008; Formica, 2003a).

Knowledge clusters shows a threefold entrepreneurial trait:

- Their organization is entrepreneurial.
- Their members seize entrepreneurial opportunities.
- Entrepreneurial patterns mold their reciprocal relations.

The Knowledge Management Cluster®

The Knowledge Management Cluster is a community of practice founded in the San Francisco Bay Area and Silicon Valley in 1998. The focus is knowledge management, enterprise collaboration, and electronic business communities. The KM Cluster sponsors and leads popular events in the Bay Area and beyond. Membership in the KM Cluster and its online community is free. The KM Cluster is vendor-agnostic and receives sponsorship from participants only.

Source: www.kmcluster.com

mation for a while, then disrupting their course and exploding into momentary confusion" (Tapscott, Lowy, and Ticoll, 1998: 89; U.S. National Commission on Entrepreneurship, 2001a). This type of behavior characterizes a knowledge pool as the outcome of spontaneous actions and serendipitous accidents as well.

Exhibit 43: Knowledge Pools

KPs are "heterogeneous groups that outperform homogeneous groups on tasks requiring creativity because of the availability of a greater variety of ideas, perspectives, and approaches to solving problems… Group members must also be willing to share their novel, controversial, or unique ideas" (Chatman et al. 1998).

Participants in knowledge pools are subject matter experts who are both individuals from unconventional backgrounds and people who come from more traditional disciplines. They are mixed together in such a way that each player comes to terms with the strengths and weaknesses of the others. Diverse conversations between people of different cultural and social background facilitate the formation of new ideas and prevent the sharing of the same knowledge. There is not a leader who holds absolute authority. The dominant configuration resembles flexible teams as in soccer and tennis doubles or in a jazz orchestra where each player performs a specific but flexible role (Fig. 9).

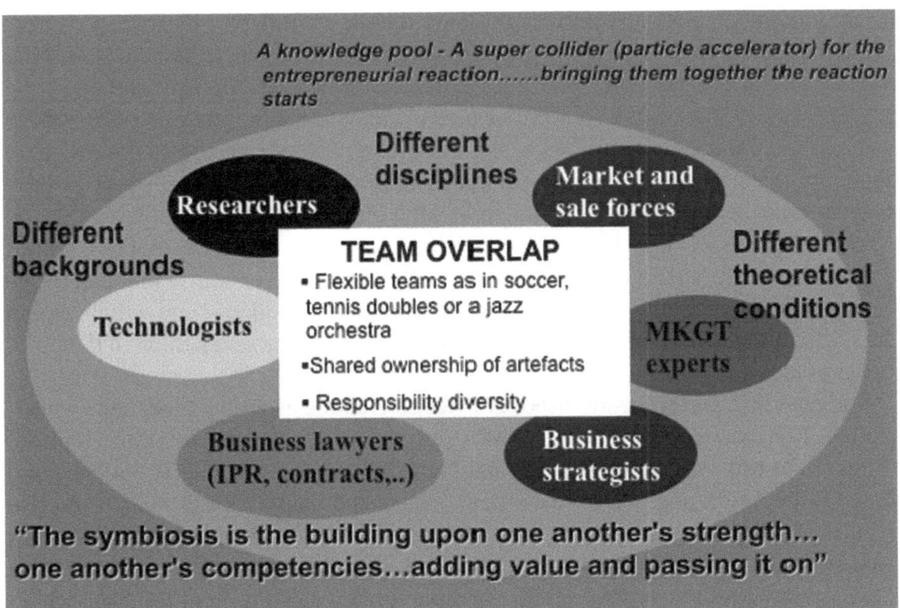

Fig. 9 Knowledge pool configuration

Knowledge clusters are strongly committed to invest in human capital as a primary source of innovation and renewal. Investments are directed to develop the individuals' emotional and technical intelligence. The search for networking and socialization inclines knowledge clusters to be focused on how people manage themselves and their relationships with others.

In knowledge clusters, knowledge sharing that occurs horizontally through the molding of communities of practice helps to cultivate an entrepreneurial culture in the research environment, to unleash business opportunities from new ideas, and to enhance researchers' productivity. Indeed, knowledge clusters serve the broader purpose of cultivating new growth-oriented ventures. Nascent entrepreneurs who focus more on growth and less on self-sufficiency can tap into the wealth of "animal spirits" and their experience inside knowledge clusters. There is a simple reason why this can happen. The opinions of the individuals who participate in knowledge clusters carry more weight than the views of the organizations to which they belong. The received wisdom of corporate principles of rational calculation is counterbalanced by the capability of the "animal spirits" to seize extroverted business ideas.

Scientists, researchers, and technologists as providers of knowledge participate in knowledge clusters to interact with users of knowledge in order to share the financial rewards of their work. Usually this happens through the formation of spin-off R&D companies. In these new ventures, which are the outcome of research, they take part as entrepreneurs in partnership with proven businesspeople, venture capital investors and supported by professional service providers.

Entrepreneurs are the engines of the cluster economy. In the past, their average educational attainment has been even lower than a secondary education. Practical experience has exposed them to the entrepreneurial process. Today an increasing number of nascent entrepreneurs hold a university degree. Their exposure to the entrepreneurial process is a mixture of formal education and field experience. New types of educational institutions – the entrepreneurial universities, described in Chap. 11 – extend the boundaries of the entrepreneurial education so as to bind stronger links between the entrepreneur and the innovation process. In Chap. 11, we expound the reasons why knowledge clusters are the kind of environment we need for creating healthy and dynamic entrepreneurial universities: those that can thrive by satisfying the demand for a better entrepreneurial education – and which can, in turn, help the nascent entrepreneur to become a more competent person in the innovation process.

Some authors have already used knowledge clusters as models for their investigations. For example, Amidon's (2002) setting of the *Innovation SuperHighway* brings to the forefront of the scene in her "innovation theatre" the practitioners who often "extend their position of leadership beyond the boundaries of their enterprise or nation." They are examples of "knowledge applied to action, [which is] the process of innovation." The knowledge cluster perspective is also discernible in Edvinsson's (2002) exploration of the "corporate longitude" side of the firm, which highlights knowledge workers who navigate "east–west" (instead of accounting analysts who navigate "north–south") so as to capture the real value of the firm, which normally does not reside in the balance sheets – to wit: "its intellectual property rights, knowledge recipes, and solutions coming from a knowledge outsider." Amidon's and Edvinsson's studies reinforce the famous assertion made by *Peter Drucker* in his *Post-Capitalist Society* (1993c): "In the knowledge society the most probable assumption and certainly the assumption on which all organisations have to conduct their affairs is that they need the knowledge worker far more than the knowledge worker needs them."

The conduciveness of knowledge clusters to innovation has two counterpoints. On the one hand, the innovative products, services, and organizations that have been incubated in knowledge clusters will form tomorrow's high-growth industry clusters. The American Entrepreneurial Revolution (Jeffry 1999) is the most vibrant example of this evolution. On the other hand, knowledge clusters innervate, nourish, and provoke innovation processes within the fabric of today's industry clusters.

Distilling wealth creation from the work of knowledge clusters requires an ability to combine production of knowledge with devices for disseminating it as widely as possible and aptitude on the part of the organizations and individuals concerned to absorb and use it.

Deep cultural roots may propitiate or prevent this from happening. Even for the most advanced industrial economies it is a hard struggle to capture the potential benefits of the modern-day knowledge society. The path toward a new age of prosperity by applying knowledge to business is full of traps. The industrial culture, mainly focused on the production of objects, does not assist in understanding the nature of knowledge, which is that of a flow and not of goods.

Notes

[1] The costs of using a market ("costs of market transacting" or "transaction costs") have been spotlighted by Ronald Coase, 1991 Nobel Laureate in Economics and a leader of the new institutional economics. His seminal works on this branch of economic theory are *The Nature of Firm* (1937) and *The Problem of Social Cost* (1960). See also *Firm, the Market, and the Law* by Coase (1990).

[2] Exemplifying this is the approach of the Global Trust Center, see http://www.globaltrustcenter.org.

Chapter 5
Embracing Business Ecosystems to Enable Sustainable and Accelerated Innovation

When the facts change I change my mind.
(reported to have said by John Maynard Keynes)

In a successful business ecosystem, the level of interconnection of products is much greater than in traditional business designs. This puts the whole concept beyond the reach of the typical corporate comfort zone. Yet increasing dependence upon an ecosystem is the only way companies will be able to continue competing in the new marketplace.

To build that dependence, and indeed interdependence, the first hurdle executives face is simply to embrace the idea – accept the ecosystem concept and work together for real results. The next step is to identify key players and determine the role of each organization in developing the product ecosystem. Changing needs call for a change in skill sets, and mapping out how those new roles will be executed is an essential part of the game plan. And finally, before putting the plan into action, it is important to determine how to construct positive, win–win relationships. In the old way of thinking, bigger players could put a squeeze on suppliers to get the lowest cost, but they would miss an opportunity to develop a relationship with those suppliers. In building a thriving, successful, and self-reliant ecosystem, leaders who have strong relationships will find themselves with the advantage. Win–lose structures are based on cost alone, but win–win relationships come from working together.

This mentality is carrying over from the business sector into government. Today, we are at the start of a heightened awareness of the power of this new way of thinking, and governments are applying the principles to create national innovation systems. For example, the governments of Singapore and Ireland are attempting to create environments that make it easy for companies to innovate. Instead of keeping strict reins on businesses, they have loosened legal requirements and allowed more flexible labor laws. This has given innovation the freedom to thrive and ecosystems the opportunity to evolve, paving the way for an entire nation to develop momentum.

In the UK, the National Endowment for Science, Technology and the Arts (NESTA) has sought to influence policy and start initiatives to make the UK more innovative. In the USA, Massachusetts is among the leaders, having developed the Massachusetts Technology Collaborative, which explores how to harness available tech sectors and create true innovation systems. In the Nordic countries, a series of fairly independent

T. Andersson et al., *Knowledge-Driven Entrepreneurship*, Innovation, Technology, and Knowledge Management 102, DOI 10.1007/978-1-4419-1188-9_5,

Exhibit 44: Business Ecosystem

A business ecosystem (e.g., the interactive multimedia industry that encompasses a variety of companies from different but convergent sectors – content, computing, and communications enterprises) is a network, or coalition, of resources, competencies, potential, energy, commitments, and promises to realize a shared profitable future. A business ecosystem often has a set of organizations collaborating – either purposefully or unintentionally – to bring forth offerings through a product or service ecosystem.

Companies in some business ecosystems are internetworked in terms of organisation, independent as regards ownership, and work "co-operatively and competitively to develop new products, satisfy a large number of customers in various market segments and incubate new businesses" (Moore 1993). In the "strategic network" model of business ecosystems, one of the companies "takes the role of 'central controller' and organises the flow of goods and information among many other independent companies, making sure that the final client gets exactly what he or she is supposed to get, in an efficient way" (Jarillo 1993).

think tanks, STEP in Norway, Fora in Denmark, and IKED in Sweden, have worked to launch and promote bottom–up initiatives in support of innovative knowledge generation and use. Such actors have enabled a more diverse approach to innovation – and at every level we are seeing the shift to ecosystem dependence. From business and industry to local and national governments, innovation is thriving through the development of strong, well-designed ecosystems. It is this new mindset that will give us continued inspiration for innovation – and the momentum to keep driving it forward.

Quickly Developing a Product Is Only Part of the Journey

In a market where timing is everything, who is not gunning for a first-place finish? Balancing the constraints of research and development with the need to quickly introduce a product to market is more critical than ever before. Speed to market itself has become part of innovation, and while we are always looking to build a better mousetrap, we also have to examine how to do it more quickly.

A sense of urgency can differ dramatically from one enterprise to the next, even though its effects are felt at every level and in every industry. The launch every 5 years or so of new hybrid seed varieties may be a red-hot introduction for a rural cotton farmer. But to a high-tech company trying to keep ahead of its competitors, the product life cycle is often measured annually.

Although speed and R&D do not always mix, the two can work in tandem. But when an organization focuses narrowly on individual product development – as most do – it misses an opportunity for innovation *and* additional revenue. The highest returns may actually come from other parts of the R&D equation, particularly when efforts are concentrated on creating a successful product ecosystem – a family of supporting products and infrastructures. While laborious in itself, a product launch should be just one segment of a thoughtfully conceived ecosystem that determines long-term success and helps enterprises thrive and *survive*.

Business Ecosystem Velocity

Like a biological community of related organisms and the environment in which they flourish, a product ecosystem comprises the interacting synergies of different entities within a company and often its suppliers, working together to bring products to the marketplace. The success of Disney's multimedia range of services, toys, games, and destinations is a prime example of a finely tuned, extended product ecosystem that propels market share.

The next layer beyond the product ecosystem is the business ecosystem. The speed of innovation or, more accurately, velocity (speed moving in a particular direction) of an ecosystem can be crucial to its sustainable success. Using the analogy of linear momentum (the product of mass and velocity), a smaller ecosystem could generate more economic momentum than a large ecosystem if its innovation velocity is faster. Often in different vertical industries, competing ecosystems try to innovate faster to capture the most value. In the semiconductor industry, there was intense competition between the Power PC (Apple, IBM and Motorola) and the Intel architecture ecosystem. Ultimately the Intel ecosystem won out, in part due to the faster rate of innovation and broader adoption of Intel architecture-based products.

In this context, the concept of open innovation – whereby an organization buys, licenses, or jointly develops new products with ecosystem partners rather than relying solely on its own research – is important. Through pooled resources, new products can be developed faster and brought to market more efficiently. Procter and Gamble's Connect and Develop strategy is a great example. The initiative connects people, ideas, and technologies to create an open innovation pipeline that is bigger and stronger than it might otherwise be. P&G's goal is to source up to 50% of new product ideas externally.

Exhibit 45: Open Innovation

Henry Chesbrough (Chesbrough 2003) has been an important catalyst in pushing the open Innovation paradigm and while he did not invent the concept, he was the first to conceptualize it. Chesbrough views open innovation as the direct opposite of the prior innovation model where vertically integrated firms initiate, research, develop, and ultimately commercialize a product from within the walls of their own organizations. Important components of Open Innovation include R&D networking, collaboration, corporate entrepreneurship (see Part II of this book), and proactive intellectual property management.

(continued)

Exhibit 45 (continued)

A key concept behind Chesbrough's theory is that through networking and collaboration, innovation can be performed faster, more efficiently, and effectively through the exchange of ideas, in effect improving the marketing of ideas to create win–win scenarios for the firms involved.

A newer version of Open Innovation, perhaps called Open Innovation 2.0 is emerging, which more broadly defines open innovation to include extended networking that allows firms to improve their innovation base so as to take advantage of societal and ecosystem capital and the creative commons at their disposal (Vallat 2009).

So how does an organization make the shift from product R&D to ecosystem R&D? How do you get your money's worth from the effort? And what warning signs should you slow down for?

Pacing the Product Ecosystem

Speed is an important factor and often translates to a faster adoption rate among users. But there has to be a balance between speed and R&D. Whilst R&D turns money into knowledge, innovation turns knowledge into money. Speed is the component that puts innovation into users' hands faster.

To achieve that with all the opportunities for additional revenue intact, it is vital that innovators focus on the R&D of a product ecosystem rather than only on the R&D of an individual product. For instance, Kellogg's has reached beyond breakfast cereals to cater to calorie-conscious consumers with a Special K product ecosystem of snacks, water supplements, and social networking groups – a horizontal approach that allows the company to meet the needs of a variety of users.

One of the reasons for this change of pace in innovation can be found in Moore's Law, named after Intel cofounder Gordon Moore, who observed in 1965 that the number of transistors that can inexpensively be placed on an integrated circuit was increasing exponentially, doubling at a rate of approximately every 2 years or less – and had been doing so since the integrated circuit was invented. The concurrent invention of the microprocessor and DRAM (dynamic random access memory) and the extraordinary economics associated with improving performance of these devices have created a wave of growth enabling more and faster innovation with 24/7 global collaboration now a daily reality.

Today that observation is joined by a second kind of Moore's Law, which says that the cost of building semiconductor factories is doubling from generation to generation. Of course, this means that semiconductor manufacturers must continue selling larger quantities of chips to remain profitable. So an examination of the typical PC ecosystem demonstrates a need for developing horizontal applications that stimulate demand and build customer value.

The creation and diffusion of technology has further accelerated both adoption and innovation. With diffusion, innovation gets replicated more rapidly. Enterprises depend upon fast diffusion, which is why the Internet has played such an important role in innovation. And, through this accelerated pace, the cost and ease of innovation has been dramatically lowered as well. For example, today's mainstream PCs and laptops deliver more power than the original Cray C1s at a tiny fraction of the cost, moving the sweet spot of innovation to individual users.

European leaders are taking a novel approach with their Living Labs project, which for example establishes small towns as laboratories to swiftly gauge and test products in a managed environment and to provide a platform for collaboration and co-innovation (see Chap. 14 on laboratory experiments). This approach is beginning to help European companies innovate at a faster rate, as they are able to see consumers' hands-on usage models and problems. Users play a major role in identifying needs, shaping applications, and even developing a more effective interaction with the developer.

Good R&D always takes time, but through the Living Labs, many European enterprises are discovering that, by involving the user earlier, they can build some of that trial-and-error into the R&D process. This allows them not only to launch products earlier and with more of the potential bugs worked out well before they arrive on the market, but also to deliver products and services that better meet users' needs and requirements.

Essentially, Living Labs provide a sterling example of the ultimate goal – to align speed to work together with R&D instead of the two constantly working against each other. The most successful regions in Europe are those that are operating as urban laboratories, and the firms located in these areas are ahead of the curve in modern technology. The goal now is to expand the Living Labs concept to a broader base by 2010 and grow this as a European platform for collaboration and innovation. A new construct/instrument developed by the European Institute of Technology and Innovation (EIT) is a knowledge and innovation community (KIC). A KIC is an explicitly designed innovative "web of excellence" which is designed to be a highly integrated partnership that integrates education, research, technology, innovation and entrepreneurship. Through an open competition, self-forming KICs can apply for funding to the EIT against various themes to help drive effective translation between, and value creation from ideas, technology, culture and business models. It is expected that the KICs will have a significant societal impact, not only through their thematic work, but also through the creation of an improved culture of innovation and entrepreneurship in Europe.

Designing an Effective Ecosystem

The collaboration between stakeholders and the various layers of an enterprise, from marketing to infrastructure, has added benefits, albeit sometimes unexpected. For example, American Airlines' parent company, AMR Corporation, made more money from Sabre, its reservation system, than it made from the actual airline in 2002. If American Airlines had focused solely on transporting customers, it would have missed out on a substantial amount of revenue. And the current poster child for product ecosystems, Apple, is an excellent example of how focusing on full integration within a product ecosystem directly affects the success of the company. If Apple had not focused on creating iTunes, and building a supporting model, it is possibly that the iPod might not be enjoying the popularity that it has today. Now, that product family has expanded beyond devices and accessories to include partnerships with companies such as BMW.

Intel's Centrino platform marketing initiative is an excellent illustration of what happens when a broader emphasis is placed beyond product innovation. Intel established an entire ecosystem to get Centrino into the marketplace, and its usability made consumers accept it. So there was the perfect blend of awareness and consumer applications. That is what successful product ecosystems need – usability, availability and, of course, room for growth. Without such utility and usability, customer adoption slows down, and these delays become the difference between leading the market and appearing as a pale imitator. A well-established ecosystem can greatly speed the public's reaction to and adoption of a particular product.

In his book *Crossing the Chasm: Marketing and Selling High-Tech Products to Mainstream Customers* Moore (1999) emphasizes the need for having everything in place before releasing a product. The entire ecosystem must work together for it to be a success: the utilities, the ease of use – from the user manual to the main features and support – all of that has to be in place. The end result is that the process will run more smoothly, and it will become infinitely easier for a consumer to adopt the product.

As important as consumer reaction is to the survival of a new product and to the longevity of its ecosystem, it is certainly not the only indicator. Good design also plays a crucial role in developing an innovative ecosystem. It is a core component that must also be looked at independently. Remember that very simple advertising slogan, "Design is a good idea"? An uncomplicated, basic thought – but it is true. You can build an entire body of work around user experience and design that incorporates packaging, interchangeable parts such as a battery charger, and consistently recognizable controls right down to the "on" button.

An interesting proof of this notion on the market today is Nintendo's Wii game console. Although other competitors seemed to have the video game market cornered, Nintendo's system targeted a broader demographic and was more user-friendly – particularly for very young and for older consumers. Now, it has expanded from a platform for video games and grown into offering products as diverse as fitness applications, which has expanded the market considerably and even drawn nongamers into the brand. It is an exemplary illustration of how great design of a product ecosystem can lead to lightning-fast consumer adoption.

In addition to design, do not forget to consider the "experience economy." You are not just selling a product, you are selling an experience – and a well-designed ecosystem offers that experience. Case in point: coffee. A convenience store sells coffee as a commodity and sees a profit – but Starbucks sells the *experience*, from the décor and in-store music, to the packaging, they are able to sell that same cup of coffee at a premium. In terms of a product ecosystem, the profit comes from creating the experience. Without a well-developed product ecosystem, that experience is difficult to create.

Some enterprises are shifting from product ecosystems to service ecosystems, attempting to develop an annuity service revenue in parallel to existing product revenue. Companies such as SAP are increasingly shifting from emphasizing product sales to including hosting deals, with the software as a service paradigm becoming a candidate for the dominant future software business model. However, this service mindset is not limited to the high-tech industry. Even manufacturers like Rolls-Royce – as referred to in Chap. 1, Exhibit 1 – are now selling "power by the hour" instead of expecting one-off payments for their aircraft engines, where advanced telemetry enables the tracking and provisioning of service by the hour.

When Speed Kills

As important as speed is, it also has many inherent dangers. The fastest car may win the race, but if it is out of control, it can be headed for disaster. A product rushed to market without proper preparation can be costly and may spell failure for the company. Remember "New Coke"? Launched in the height of the soft-drink war, Coca-Cola did not foresee that consumers would be so resentful about the withdrawal of the original formula; a subsequent reintroduction later led to great sales gains, but it was a major marketing lesson.

Yes, getting to market first is significant, but so are quality and the impact of consumer reaction. It is a thousand-fold more expensive to repair bugs or problems with a product after it is on the market than to fix them when it is still in the design and development phase. That means you have to manage speed vs. functionality – and common sense must win out every time.

It is essential to develop an index of cost, speed, and functionality. If a product only has 70% functionality, getting there first does not matter; it likely will not succeed. There has to be a compromise in the mindset of all the players – it is good to be the first to market, but it is hard to be the first to market *and* be perfect. Today's dramatic reduction in preliminary research costs and the rapid information available through the Internet mean that good ideas will be copied almost instantaneously by competitors. So, timing and quality count.

Consumer electronics offer many great examples of the fast pace of innovation – and offer further proof that being first does not always guarantee success. Apple was not the first MP3 player to hit the market, but it benefited from other players that appeared on the market early and had problems. Apple improved upon those ideas and became the undisputed leader in that arena.

Thus, the balance of innovation and common sense also requires that leaders shift their skill sets and become more entrepreneurial in nature. For the past 50 or 60 years, business leadership has been about efficiency and managing costs – cutting expenses and bringing the lowest cost product to market. Today, it is all about personalization for the consumer and, because of that, leadership means developing more intrapreneurial and entrepreneurial skills – the ability to innovate and shift not only your focus, but your resources as well.

Information Technology (IT): A Critical Factor

One of speed's critical allies is IT. We hear a lot of talk about our 24/7, always-on world, and IT is what makes that possible. Now emerging are collaborative design tools that make it easier to innovate and will increase the speed to market on a global level. Managing IT and innovation is a crucial part of the ecosystem equation. More than ever before, timing is becoming another of the most critical components of success. The adoption of rapid solution prototyping can help enterprises improve their innovation capabilities and cut down the amount of time it takes to get a product to market.

Of course, there is a tremendous difference between consumer adoption and business adoption. While consumer adoption may require the customer to adapt or discard one or two items, in the business realm the changes become much more significant. Adoption of a single new product can trigger a change that affects tens or tens of thousands of other things. Every new device introduced or adopted by a business has to work with the current system, and that integration can impact a lot of equipment and staff. Accordingly, the need to have an efficient product ecosystem in place is exponentially more important when dealing with business adopters versus consumer adopters.

Still, carefully integrating and managing IT and innovation can have a big payoff. Nowhere is that more clear than in Formula One. In that world, of course, speed is crucial because the fastest car wins the race. However, innovation is also imperative. So teams now use a combination of technology and innovation to increase the speed of their cars. Supercomputers and design teams mock up new designs and then use computational fluid dynamics (CFD[LT1]) to see which ones work best. From that, the design team fabricates two or three models for testing in wind tunnels. This approach greatly increases the speed of innovation through computer simulation, and it is just one example of the impact of IT. Innovation is no longer viewed as an act of serendipity: it has become a necessary element that must be managed and nurtured.

The proper use of innovation can create the momentum needed for an ecosystem to succeed. As we bring new products to market, innovation enables us to turn unlimited possibilities into valued realities. And in a well-planned, well-managed environment, that ecosystem momentum provides companies with the speed they need to succeed.

Chapter 6
Corporate Management of Knowledge*

You never change something by fighting the existing reality.
To change something, build a new model that makes
the existing model obsolete.

(Buckminster Fuller)

Corporate management of knowledge (CMK) aims to gather, create, and share knowledge in order to promote organizational innovation that delivers value to the ecosystem of partners, customers, employees, alliances, and other relationships in which the company is involved. There must be a high level of sensitivity toward the interdependence with all those players.

CMK looks at knowledge from the organization's point of view. In particular, the focus is on organizational innovation that improves organization's processes as derived from the organization's business objectives. Contents, contexts, and communities (the "triple C") capture the essence of this approach.

For the purpose of CMK initiatives and processes, knowledge is categorized as highlighted in Exhibit 46.

The Role of Chief Knowledge Officer

Corporate judgments are affected by the rate of change in the external and internal business environment, and many decisions are made with only a limited amount of the necessary information. Moreover, managers in their own isolation justify systems leading to knowledge silo.

The chief knowledge officer (CKO) is positioned where the action takes place but must create the impression that he/she is going to improve business processes (i.e., less redundancy and more efficiency and effectiveness) and is not inquiring to find faults. By asking questions that lead to conversation, the KCO contributes

* This chapter draws from a body of experiences learned at seminars and workshops in which the authors have taken part, as well as unpublished papers.

T. Andersson et al., *Knowledge-Driven Entrepreneurship*, Innovation, Technology, and Knowledge Management 102, DOI 10.1007/978-1-4419-1188-9_6,

Exhibit 46: Corporate Management of Knowledge: Categories

1. Category for latent knowledge:
   Elicitation. Explication of unarticulated latent knowledge or engendering new insight(s).
   If successfully performed, the knowledge in point becomes explicit.

2. Categories for explicit knowledge:
   Codification. Articulation and transit of explicit knowledge from a human source to any kind of media, either straightforward (e.g., plain text or model) or adapted (e.g., embedded in a work procedure).

Once codified, the knowledge is detached from its source and is transferable to others. Subcategories are as follows:

- *Embedding.* Codifying (hard-coding) knowledge into machines or software applications, thus annulling any discretion whether to apply that knowledge.
- *Prescribing.* Codifying knowledge in a strictly guiding manner ("how-to"), like manual, procedure, rule, prescription, recipe, etc., which makes it applicable unbeknown to the background rationale. Still (unlike embedding) the knowledge's application is subject to a human's compliance.
- *Structuring.* Here, the metaknowledge (i.e., knowledge about knowledge) is the object, rather than the knowledge itself. By framing, classifying, ordering, and interrelating concepts and/or terms a knowledge domain is constructed under a shared language.
- *Abstracting.* Codifying knowledge in a form of theory, idea, reason, etc. that answers the question "why," thus interpretation and modification are required for its usage.
- *Detection.* Intended or accidental identification of required explicit knowledge.
- *Assessment.* Entitling explicit knowledge with credibility, value, significance, or meaning, either actively or by omission (e.g., ignorance, unawareness).
- *Transfer of knowledge.* Addressed transit of explicit knowledge from a human source directly to other human(s).

3. Categories for tacit knowledge
   Assuming that tacit knowledge is inexplicable, the tacit realm is tightly delimited, allowing just two options:

- *Transfer of knowledge holder.* Making tacit knowledge available by repositioning its source (individual or item that embodies the knowledge).
- *Nurturing.* Assisted recreation of tacit knowledge inside another person than the holder.

to knowledge creation. In relationships with managers and employees, the CKO has to be effective in mastering the information realm (what he/she is able to convey about what he/she knows) and the knowledge realm (his/her capability of interpreting information).

The CKO figures out how gathering knowledge across the company's value chain will unfold and shows managers and employees how they can be part of a framework of a comprehensive four-phase life cycle of knowledge: capturing and documenting, retrieving for reuse, creating, and sharing knowledge by taking care of the knowledge flow (see Appendix 8 as to the influence of eight dimensions on knowledge sharing).

Exhibit 47: Chief Knowledge Officer

The chief knowledge officer is a facilitator who is responsible for brokering connections between seekers and sharers of knowledge.

There is a critical point, namely to match the right kind of "seeker" with the right kind of "sharer." Some valuable experts may not be willing to use PCs and e-mail for sharing, so they are only useful for face-to-face consultations. Some may not want to talk to apprentice-level people, whereas other experts may like that kind of mentoring role.

The knowledge facilitator creates nodes (detecting the individual actors within the company's networks) and ties (detecting the relationships between the actors).

Doing his/her job well requires an understanding of these very human and perhaps unquantifiable dimensions. With the right mix of people, and the right kind of interaction, something magical happens. But it will probably always remain a mix of art and science.

The CKO needs also to rely upon the following:

- Stories generated by the business on how the services provided and solutions deployed have been beneficial.
- Listening to customers and providing the services and solutions that they ask for.
- Adjusting activity yearly to meet changing business needs.
- Constantly marketing and selling his/her program to senior managers inside and even outside the company.
- Jointly developing annual goals with internal customers of knowledge management and regularly tracking progress toward milestones.

The CKO takes care of three levels of qualitative measure:

- Performance measures (measuring the act of sharing; who and with what).
- Throughput measures (measuring the outcome of this action, as if and by whom it was retrieved for further reuse).
- Operational or business results measures (for cases where the reusing agent could identify a tangible advantage gained from this action).

For this purpose, the CKO identifies and selects company employees who are ready to be inserted in a framework of collaborative thinking. These are the company's "knowledge champions" (KCs).

Exhibit 48: Collaborative Thinking

The essence is to find people in the same field across the company, help them to admit that they face technological or organizational or market difficulties in implementing different aspects of the same technology, and suggest that maybe they could gain by conferring with a colleague from a different division. All together, they give shape to a community of knowledge practice.

The company's interest in it comes from the realization that if the employees personally solve their problems it would enhance their performance in divisional projects. The value generation for the company is therefore subtle and kept below the cognitive level of the community, in order not to awake sentiments of business impact that would put the participants in competing and opposing parties.

Exhibit 49: Knowledge Champions

Knowledge champions are natural thought leaders who assume their role within their business unit and use their leadership position to create a collaborative culture that recognizes, quantifies, and communicates value.

One of KCs' central virtues is their ability to leverage the knowledge of others, especially their strengths in perception, diagnosis, and prescription. The process of asking "Is there anything else I need to know," which is triggered by the CKO together with his KCs, is expected to educe in other managers and employees insights that they later say they were not aware of. In other words, managers and employees will feel themselves less constrained by people's typical attitude of being more willing to display knowledge than ignorance.

The CKO as a Seeker Rather than a Knower

The CKO's ability to be comfortable with discovering what he or she does not know is a learned and cultivated behavior. The CKO deliberately exposes himself/herself to different situations, cultures, domains, and groups where he or she has poor knowledge or little previous exposure. In doing so, the CKO manages his own identity to be a seeker rather than a knower – a seeker who:

- Is curious – that is, engaged in continual inquiry. In the interaction between the CKO as "asker" and the managers and employees as "askees," information passes in both directions, through the answer as well as through the question. "Listen to the patient; he/she is telling you the diagnosis" – listening is part of the art of "silently" asking.
- Tests people by exploring if he/she could have a good conversation with them in a coffee shop or driving from A to B.
- Participates in face-to-face dialogues and online conversations in sparse networks, leveraging knowledge champions.
- Discovers the common areas of cooperation/rivalry between different corporate departments.
- Breaks established personal mental models (process of "creative abrasion"), so the corporate culture can be made to evolve toward collaborative performance.
- Releases tacit knowledge through trust building and knowledge dynamics – which makes expressible what has previously been unexpressed or implicit knowledge.

The CKO Designs Conceptual Maps to Discern Creativity in the Corporate Environment

Creativity is the essence of corporate growth and relates to the generation of new ideas that are converted into economic activity (as shown in Chap. 10).

To discern creativity, the CKO designs conceptual maps (see Fig. 16 in Chap. 10) that help managers and employees who encounter knowledge outside their organization to recognize and benefit from it.

The CKO takes the initiative needed first to collect knowledge and then to incorporate it into the Intranet. He also takes responsibility for relating the external acquired knowledge to whoever would be interested. This would actually generate new knowledge for the organization. When this happens, the "not-invented-here" syndrome has been vanquished.

"Sparring" Is a Key Method of Knowledge Disclosure and Discovery

The CKO accomplishes the task of seeking common ground when getting people to know each other and building trust. Yet once that is accomplished, most progress is based on the ground that is not held in common. This means that the CKO has to search for both similarity and diversity. The CKO motto has to be: "Connect on people similarity (i.e., common ground) and profit from their diversity."

"Co-operation and harmony are both desirable ends... However excessive harmony becomes soporific. Critical faculties are lost and innovation disappears" (Hunt 1998). Therefore, within a frame of social cohesion, the KCO must enhance dialogues and discussions that raise cognitive conflicts.

Exhibit 50: Cognitive Conflicts

These conflicts in the form of debates about strategies, issues, solutions, and appointments are constructive and energetic rather than destructive and emotional like affective conflicts.

Intellectual disagreement stimulates new ideas and puts in motion a process of idea collision. This produces a real quantum leap – new knowledge, unusual insights and features (Leonard 1998).

Through this process, the personal knowledge of managers and employees turns into organizational knowledge.

In this respect, the CKO should discover questions that are inconvenient to deal with (i.e., questions that expose uncomfortable points), so that through the consequent debate and dissent he can detect weak signals (i.e., seeing things early enough to do something about them).

The CKO as a Cross-Pollinator for New Initiative Processes

The CKO contributes as a broker or cross-pollinator to the formation of integrated product/service/venture project teams inclusive of individuals with unlikely backgrounds – technical and nontechnical people from the various interested corporate departments, and enriched with strategic suppliers. Each team creates data and information in the process of advancing the specific project toward the next development phase.

These teams should resemble a soccer team, a double tennis team, a classic orchestra, and a jazz band where each player covers a specific but not rigid role and each player comes to terms with strengths and weaknesses of all the other players.

From Subjective Behaviors to Quantifiable Results

So far, we have expounded how the CKO can set in motion a satisfactory process for sharing and creating knowledge. The focus has been on the behaviors of all those persons, primarily the CKO and his knowledge champions, involved in knowledge management.

Equally relevant is the identification of intellectual capital indicators and the development of an intellectual capital report. From this perspective, the focus is on quantifiable results (in terms of growth, renewal, efficiency, and stability/risk) that can be achieved rather than on subjective behaviors (see Fig. 10).

The CKO takes a gradual approach to the building of intellectual capital indicators.

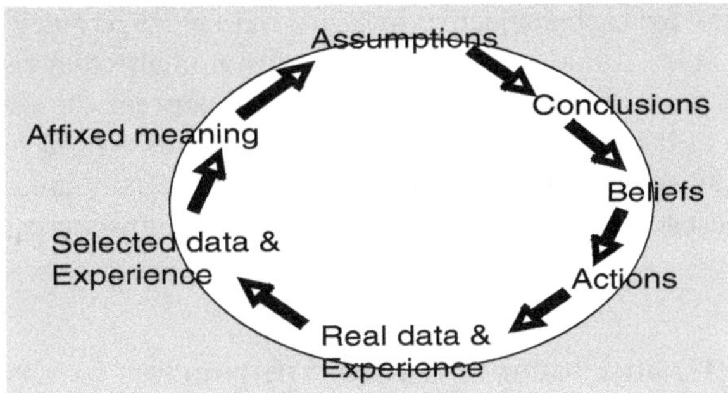

Fig. 10 Subjective behaviors and quantifiable results: a virtuous cycle

All of the quantitative methods have their criticisms. These generally focus on the inability to properly capture the value of intangibles: the reason for which it seems more appropriate to take a course of action that allows for a gradual approach to the indicators and stepwise refinements rather than buying tools for intellectual capital indicators made available by several firms specialized in knowledge consultancy.

It is important that the CKO and his/her KC team understand and agree which forms of intellectual capital are most important to the organizations and which forms of intellectual capital will drive sustainable performance (see Appendix 9).

Monetary and Subjective Measures

There are several limitations to monetary evaluation, namely:

- Many intangible assets do not have market prices because of their tacit and unique characteristics.
- Most intangibles are not sufficiently traded to determine a comparable market value and it is often impossible to disassociate them from tangible assets.
- Intangible assets are dynamic. In reality, the value of such assets is only realized when they work together in complex and fluid relationships.

Monetary measures are, for instance, applicable to R&D expenditures, spending on training and patents, and copyrights (valued via future cash flows discounted by a discount factor). Other indicators, like the number of researchers, can be calculated in absolute numbers.

Subjective measures using rates between 1 and 5 are applicable, for instance, to employee satisfaction ("Is working here a pleasant experience?") and customer satisfaction.

The weightings for each subjective measure can be dangerous if the CKO gets it wrong, as the index would not fully reflect a given intellectual capital asset. On the other hand, a good dialogue about the weighting between the CKO and his/her team can be one of the most valuable contributions of this approach.

Detecting weak signals and profiting from people diversity are actions that the CKO has to undertake in order to reduce the risk of data misinterpretation.

Cost-, Market-, and Income-Based Approaches to Evaluation

The cost-based approach determines the value of an asset by ascertaining its replacement costs. Replacement costs involve fewer practical limitations than historical costs, but it is very difficult to assess the full replacement cost of an intangible asset because it is hard to separate the intangible asset from other assets.

The other difficulty with the cost methods of valuation is that there is an implicit assumption that expenditures always create value, which is not the case.

The market-based approach determines the value of an asset by obtaining a consensus of what others in the market have valued the asset at.

Market methods estimate the market value of an intangible asset by comparing it to similar intangibles that have been licensed or sold recently.

The income-based approach determines the value of an asset by looking at the income-producing capability of the asset.

The difficulty with income methods is isolating the income attributable to an intangible asset. In using income projections, income generation must be allocated between the intangible and associated tangible asset. Estimating the expected remaining life of an intangible asset is also difficult, but is fundamental to the calculation because income methods are sensitive to variations in remaining useful life where the life estimate is below 20 years. Other difficulties include choosing an appropriate capitalization rate.

The CKO Operates in a World of Plausibility

There is an exaggerated tendency to confuse statements and data that claim to be facts with those that assert a desired behavior. Moreover, on a personal and subjective basis, "facts" may have different significances depending on one's point of view. Hence, the CKO is expected to be coherent about the constraints under which he/she has to operate and to search for correspondence or dissonance between facts and believed phenomena.

Chapter 7
Knowledge-Relevant Economic Policy: Analyzing Knowledge Policymaking in Managed and Free-Market Economies

We've opened the minds of colleagues; now we must do the same for policymakers.

(Thomas F. Malone)

Theories, when they have gained credence, become vested interests. The prestige and livelihood of schools and teachers are bound up in them; they tend toward enclosed doctrine, not open to fresh information.

(Isabel Paterson)

Although the power of knowledge to forge the economy and society has never been so memorably and sharply delineated as in current circumstances, most companies have not succeeded, so far, in seizing the vast opportunities available for knowledge innovation. In an ever more demanding market of talents, knowledge capacity is underutilized and industry leaders do not aspire to follow in the talents' footsteps.

This disheartening economic climate depends on how modern economic policy is shaped. To take a course appropriate to the "knowledgefication" era, the flow of energy needs a proper policy. The distinguished novelist Isabel Paterson laid the foundations of the reasoning behind this recommendation in her masterpiece, *The God of the Machine*, published in 1943.

Sixty years or so later, the time had come to reappraise that theme. In their celebrated book, *The Knowledge Creating Company*, Nonaka and Takeuchi (1995) portrayed the Paterson's flow of energy as a flow of knowledge (see Chap. 2).

Today, Nonaka's and Takeuchi's view is entrenched in managers' thinking. By far less explored and even more controversial the argument seems to be that the organizational form and its evolution (organizational innovations) at the flow level are under the influence of the political institutions' workings and, therefore, the government policy through regulation on knowledge creation, dissemination, application, and management.

Indeed, the government policy stance should take something of a front seat in the knowledge economy. Knowledge flows with contained context and values.

T. Andersson et al., *Knowledge-Driven Entrepreneurship*, Innovation, Technology, and Knowledge Management 102, DOI 10.1007/978-1-4419-1188-9_7,

Knowledge-relevant economic policy (hereafter "knowledge policy") carries far-reaching implications on both.

This chapter attempts to throw the spotlight on the nature of government policy as a mechanism, pitting two policy models against each other: namely, the state-sanctioned corporatist, "closed shop" model, which results in a "closed economy," against the anticorporatist, "open shop" ("free economy"). The former is molded to the body of a compartmentalized society with a strong bias toward narrowly based, special-interest groups whose members place their interests before those of the market as a whole. The latter is designed to fit into the wide-open spaces of a knowledge communities-oriented society.

Exhibit 51: Underutilization of Knowledge Capacity

Albert Hochleitner, director general of Siemens, Österrich (Austria) said at ALPBACH in 1999:

"Less than 2 per cent is spent on the future perspective. Some companies are even lower than 1 per cent! Although the urgent business of everyday life is important, it is not as important as the future.

Companies only use 40 per cent of the knowledge of their staff. Many other knowledge experts would consider this a very modest figure.

Knowledge has to be identified; knowledge must be distributed; and new knowledge must be acquired."

Source: Report of the ALPBACH 2000 conference (http://www.entovation.com/gkp/alpach.htm).

Exhibit 52: Knowledge Policy

A sound knowledge policy incorporates market-oriented measures that set the stage for good economic performance in the knowledge economy.

These measures include, inter alia:

- Releasing individual talents.
- Protecting the independence of change agents and minimizing the risk of "capture" by specific business interests.
- Putting competition back on the road to transparency (instead of secrecy-set and restrictive practices).
- Contributing to the practical application of knowledge.
- Promoting a vibrant social capital infrastructure founded on the intelligence and interactions of people in voluntary associations.

A critical atmosphere surrounds knowledge policy. In Europe, there is condescension toward the intrusive hand of government regulation, rooted in the tradition of tripartite corporatist arrangements involving labor (trade union confederations), capital (employer's confederations and trade associations), and government.

Investments in the political market made in the past for the purpose of empowering those arrangements have created "path dependence."

The increasing involvement of regulators does not bode well for the knowledge economy. Policymakers first fail to see the relevance of knowledge innovation agents and are also skeptical about sound actions by them. They therefore prefer a safety first approach based on the precautionary principle of a corporatist type of policy that shows little appetite for growth-promoting reforms, without which advancements of the knowledge economy are doomed to be stifled. What is more, much-needed government plans to loosen laws and regulation protecting professional guilds and corporatist groups from competition are likely to be smothered by muscular, home-state corporate giants, for regulation can be a useful barrier to competition.

The investment required to establish a new policy course is an adjustment cost made more onerous by the bureaucracy of corporatism, which is merely a means of maintaining the status quo. This is the reason why gradualism has unceasingly been used to justify much-needed structural reforms in the education and business foundations on which human capital has been built during the machine age. The outcome, unfortunately, is a reform paralysis in which the chokehold of vested interests plays a big part. But the forces at work are more complex. This paralysis can also be attributable to the actions of policymakers.

The greatest challenge that society and firms face in the knowledge economy age is to connect meaningfully with knowledge innovation agents to enhance their capacity to handle innovation as the process of putting knowledge into action. The economic policy that complies with this challenge contends that free makers of knowledge innovation and free knowledge markets are corollaries, and hence it embraces the principle of a light, free market-supporting regulation that prevents corporatist forces from disrupting the autonomous operation of the free market.

The corporatist attitude, which cuts the line of energy and therefore constrains knowledge activity at many levels, raises an important question: can knowledge innovation agents' entrepreneurism exist freely, or would their freedom of action be curbed by corporatist groups strong enough to impose controls and restrictions upon their industry?

Exhibit 53: Knowledge Innovation Agents

These are free agents – makers of knowledge innovation – who operate in competitive markets that are an expression of personal freedom.

The concept of "free agent" is the counterpart to the "citizen," with its corresponding demand for participation not mediated by corporatist institutions to which the individual citizens ought presumably to belong.

The free agent is committed to the wider view and long-term shared undertakings for innovative projects that can tap into the wealth of knowledge and experience of other agents, in order to promote economic growth along with wellness, social development, and collective solidarity of a self-regulated community (Formica 2003a: 265–268).

(continued)

Exhibit 53 (continued)

In a corporatist society, free agents flourish on the edges of the corporatist groups, far from the influence of their rigid rules and strictures that impede the emergence of "different" ideas.

Free agents hold the view that the term "competition" refers to the energy needed to win and do better for oneself and as part of a team. They also acknowledge that promoting competition intersects with promoting innovation.

Free agents are cross-functional disruptive individuals belonging to multiple communities of knowledge practice in which hierarchies are flattened and boundaries made permeable. Unlike professional corporations, they are organized across industries, sectors, and functions.

Free agents coalesce in a "small, informal team of creative people and informal networks of creative groups and communities of practice unfettered by chains of command, free to exchange insights and inspiration. They tend to be the key players in defining radical innovation projects" (Community Intelligence, white paper, May 21, 2003).

In other words, free agents are loosely linked professionals whose independence of mind ("independent creativity") and openness to change ("interdependent cooperation") are unconstrained by conventional corporatist categories.

Their collaborative governance is a kind of governance that is loose, informal, undeclared and allows for a number of free-agent communities to work together toward a greater good.

Free agents are business-sensitive doers who negotiate on a performance contract basis with clients in the marketplace. This leads to collaborative projects between them and mutual disclosure of secrets.

It is worth noting here that the guild regime forbade a worker who possessed a professional secret to divulge it once he was in a foreign territory. We have inherited our hostility to the idea of disclosure.

Today's professional corporations fear transparency because the opening of their iron curtains for the purpose of making exchanges of knowledge with outsiders would tip the balance of power toward breakaway institutions such as the communities of knowledge innovation agents.

Corporatism Takes Centre Stage

Corporatism hurts the knowledge economy, for the corporatist structure breaks the rules of conduct that would allow maximum advantage to be reaped from the flow of knowledge. It prevents adjustment to changing circumstances and is therefore ill-suited to contexts of high innovation; there is a Luddite reaction to knowledge progress made by free agents of knowledge innovation. It is the corporatist power that takes the lead in shaping knowledge policy to its liking.

The blade of the policymakers' sword lies on the knowledge workers' neck, while policymakers have an interest in representing the pleas of corporatist groups.

In a knowledge economy who are the people who should replace the leading corporatist actors of the national-industrial state? Answering this question implies rethinking the policy model founded on the exercise of power through the bureaucratic structures of corporatist organizations. Voluntary and open market-based communities of knowledge practice should be overtaking the corporatist groups that operate in sheltered markets. The former outperform the latter because their flexible processes of disciplined pluralism promote innovation (as opposed to the rigid processes of forced allegiance to the guild). The philosophy of communities evolves as a set of practices that gradually become established through trial and error as expected patterns of knowledge innovation.

The fundamental political problem of knowledge-led economic growth is thus to devise the appropriate means for channeling government action into support for CKP-driven markets, so predatory behavior by professional corporations (the fiefdoms of modern times) would be totally alien to the policymakers' collective experience.

Oddly, knowledge policy is destined to become a leading villain, both in government and across industry. A growing queue of free agents of knowledge innovation is petitioning the government for relief from corporatist policy intervention, while corporatist groups have been successfully seeking access to policymakers' offices to exert and gain special privileges.

Corporatist group cognoscenti, led by media pundits and academics, do not recognize that the corporatist model has repeatedly failed to rise to the challenge of the knowledge economy. Instead, they see the corporatist government as impresario of the knowledge economy.

Exhibit 54: Corporatism

"A system of interest representation in which the constituent units are organised into a limited number of singular, non-competitive, hierarchically ordered and functionally differentiated categories, recognised or licensed (if not created) by the state and granted a deliberate representational monopoly within their respective categories in exchange for observing certain controls on their selection of leaders and articulation of demand supports" (Schimitter 1979: 13, quoted from Lloyd-Jones 1994: 17).

Corporatism places undue importance on the stakes of the state-sanctioned categories, which are entitled to define the public interest, as against the interests of other societal goods such as the state-unaffiliated, voluntary associations of knowledge innovation agents.

Corporatism is a world of separate experiences within which innovation is constrained by the corporation's code and assumptions about how to do business and negotiating relationships. Power flows through negotiations of corporatist leaders with government officials and party cadres. A better breeding ground for conformity is hard to imagine.

(continued)

Exhibit 54 (continued)

Corporatism shows a strong inclination to set everything in rules and regulations – which results in a bloated bureaucracy.

The corporatist government is the one whose economic activities are not distinct from the economic interests of those who control it.

Corporatist states are the most rule-bound countries.

Corporatist groups hold the contrasting view that changes would damage their status and power. They try to fight change and accordingly rely on the embodied principles of slow-moving times. Arm in arm with government rules, they take decisions about how professionals must be locked into particular roles (including the compulsory association that bars outsiders and unknowns from practicing the profession or keeps them out of the loop), how members of professional corporations operate, how they set prices, and even where they can sell their services.

Up to now, corporatism has guaranteed the status-quo security, which is the counterpart of the social solidarity that laid the foundations for the welfare state.

In the 1980s, many of Europe's political leaders, political scientists, and economists sought to restore the concept of corporatism as a means of governing the domestic economy (Andersson 2000: 1). In those days, MIT's Lester Thurow argued that corporatist, consensus-based, and coordinated market economies in Europe were "best placed to excel in the new, knowledge-based industries thanks to [their] tradition of corporatism and [their] publicly financed intellectual infrastructure." By contrast, liberal market economies, which rely on free market mechanism to regulate economic relations, were unable to do their job properly. The US economy, for example, "was too competitive and cowboyish for that." All this was said and written at a time when the US economy was to become the most attractive place for the advent of the tech boom (Thurow 1992; Watson 2002: 15).

Policies built on corporatist group–state relations are not suitable for current conditions prevailing in the world today. Corporatist countries are on secular growth downtrends.

As it has been noted by Andersson (2000): 1), "corporatism could be a successful social organisation in a stable economic environment, but in a volatile economic environment this need not be the case." In the consensually managed economies, where the intellectual and corporatist elites dread the restructuring that the free-economy and free-agent model brings in its path, the time is ripe for a policy regime change that should unleash the flow of knowledge still hostage to the influence of vested interest groups. This policy therefore aims at curbing the corporatist groups' power instead of meddling with them. It also makes clear that these groups can no longer rely on numerous central and local government subsidies to protect and promote vested interests.

Exhibit 55: Coordinate Market Economies and Liberal Market Economies

Coordinate market economies (defined also as mixed, consensually managed, or socialized economies) are characterized by corporatist tripartite arrangements, which encompass government, business organizations, and trade unions.

At the nation-state level, and particularly in Europe, this type of arrangements is the archetype of corporatism and has a long history.

"Co-ordinate market economies have a set of interlocking policies and institutions, which include long-term relationships between companies and providers of capital, extensive collaboration between employers in the same industry, especially in standard-setting, vocational training and pre-competitive research, and a commitment by companies to equip their workers with firm-specific and industry-specific skills. Liberal market economies rely on market mechanism to regulate relations between companies and their workers, between users and providers of capital and between companies and their suppliers" (Hall and Soskice 2001)

In a market-oriented liberal economy the policy approach is for less and less government intervention.

The locus of access to knowledge markets should move from corporatist, interest-group representation to free agents of knowledge innovation, who need the level playing field of the liberal market economy. These agents should rely on free market-supporting regulation, which would help them to cope with volatility and rapid changes of the economy by playing to their strengths. This type of regulation acknowledges that the public interest content of knowledge can be achieved, lending force to the expression of personal freedom and competencies. Conversely, any sort of policy that conjures with the corporatist groups' views of regulation ends up relying on government handouts that subdue a free, truly open market.

Springboards for Corporatist Policymaking

Corporatist policy deliberations are the product of bargaining and consensus – that is, of exchange and ongoing negotiations between corporatist groups, syndicates (the so-called "social partners"), and government within a developed corporatist institutional structure. There are formal institutionalized relations between these organized interests and government policy.

Special-interest groups, whose access to public authorities makes it possible to influence policymakers to fit specific interests, easily sway government policy.

"Double-positions" – the fact that representatives of corporatist institutions often play an important role in political parties – help to align corporatist and public interests (Saul 1995).

The inspiration of corporatist policymaking lays partly in principles drawn from the medieval system of craft and trade guilds that brings overregulation, politicized practices, barriers to competition, and lack of transparency: in other words, blockages to the flow of knowledge.

For most of the nineteenth century or so, a corpus of laws was drawn up to establish "officially recognized knowledge" of practice. Policymakers undertook the consequent action of protecting by means of licenses the roles assigned to the professional corporations stemming from that legal framework (Dingwall 1996: 4).

In modern times, the biggest thrust to this enduring process of corporatism-building came from the communitarian and corporatist theories of the 1920s and 1930s: on the one hand, the "state corporatism" of Primo de Rivera and Mussolini;[1] on the other, the "societal corporatism" of Wilson's National Economic Development Council and Roosevelt's New Deal (Lloyd-Jones 1994: 4; Formica 2003c: 243–244).

Exhibit 56: Guilds

In medieval times, merchant guilds and craft guilds were exclusive, regimented organizations in the form of specially organized groups whose essential purposes were as follows:

- Control of economic life through the creation of monopolies.
- Preservation of monopolies through limits on the number of masters and the interdiction of disclosing information to outsiders.
- Closed shop: to practice a specific art or trade in a town one had to be member of the appropriate guild.
- Restraining the initiative of the more enterprising.
- Permitting innovation only if all producers could share alike in the benefits.
- Regulating all prices and forbidding price-cutting.
- Banning many kinds of advertising in order to regulate competition between members.

The aim of each guild was not only to establish an effective monopoly of a given profession, but also to monopolize competition for it. No one had a right to do what he liked, but everything was managed in the best possible way according to the professional code.

Legal constraints and the legacy of tradition mean that today's national professions are guilds reluctant to embark on the process of becoming less restricted or limited in their scope. The desire of each profession to remain aloof from the rest holds back mutual interactions at home and, worse, internationally. Professional corporations prefer speaking to themselves through their introverted (self-interest) vertical hierarchies instead of harnessing complementary competencies through extroverted (external) collaborations with colleagues of another practice.

If dictatorial governments like Mussolini's fascists in Italy gave legislative representation to corporatist groups, the leap forward into corporatism under the New Deal saw policymakers and administrators come under the influence or even domination of special interest groups.

From then on, as in the Middle Ages, there has been a close connection between the public authorities and the representatives of those groups. This connection is twofold: on the one hand, corporatist officials are appointed to serve in local and national governments; on the other hand, for whole classes of professional corporations (e.g., lawyers, doctors, consultants) government is both a vital funding source and the main entity with which they work helping organizations and individuals to deal with it (Locke 2002).

Knowledge Policy in the Public Interest: Plenty of Room for Policymakers?

The frustration arising from the corporatist model has become apparent with the growth of the knowledge economy. Yet, the clock is still turned back by corporatist policymakers who wish to remold today's knowledge society into a likeness of the medieval guilds. The more policymakers' workings are opened to corporatist consensus, the more effectively lobbyists and special pleaders can manipulate them.

The government has to "take responsibility" and "do something." This widespread habit has gone further than the management of the flow of material goods. With the advent of the knowledge economy, the stage has been set for ensuring the flow of knowledge and intangible assets (e.g., brands, patents, R&D projects, trained staff) by means of policy measures.

Accordingly, policymakers step into the role of a context builder that should favor human interactions aimed at transforming those assets into wealth-creating resources. The prevailing doctrine, stated simply, is that policymakers have to deploy and direct resources toward intellectual capital with a view to exercising more effective sway over individuals and organizations that create, disseminate, and transform knowledge.

Here, the question to focus on is under what conditions the market offers the natural and inherently most favorable milieu where "free" agents are allowed full freedom in making their decisions on innovation. Naturally, there are many cases in which that is not necessarily the case. Under certain conditions, proper markets may simply not exist [e.g., when it comes to our global environmental assets such as (depleted) fish stocks, tropical forests, or air or water quality, it is far from trivial how orderly market conditions can evolve]. We have noted the poor state of intellectual property rights. We have noted how a lack of accountability has played a crucial role in the squandering of billions of US dollars over the last decade and in bringing world financial markets to their knees.

In many such cases, the underlying reason, however, has to do with institutional failure – for example, the failure on the part of government, or multiple governments in collaboration, to assign or allow for the formulation of property rights, or failure to institute orderly conditions for the development and diffusion of appropriate information. The latter may result in genuine uncertainty in the marketplace, or the presence of costly asymmetries in the availability of information, causing the better informed to exploit the less well informed. In such cases, the appropriate policy response has to do with rectifying those sources of failure. It is a different thing to argue, however, that decisions should simply be removed from market actors and be deferred to someone else – "letting loose" the functional representation of the corporatist system in ways that thoroughly disrupt the autonomous operation of the free market.

Put differently, the choice is between:

(a) "Constructive" government policy that encourages brain circulation and helps to underpin orderly market conditions, so that real needs and real issues can be taken into account in the marketplace.
(b) "Restrictive" policy that serves to limit people's ability – as individuals or in their role as agents of private entities – from developing or using knowledge in new ways, through a patchwork of rules and procedures.

In most instances, the former approach doubtless offers the best way to generate a favorable exchange of knowledge and enable a myriad of agents to embark on innovation journeys whose objectives are not predestined but which can be colored and redirected through continuous contributions of human creativity and innovation.

The second approach, on the other hand, means that the state and professional incumbent corporations will exert a strong influence on the dissemination and use of knowledge. That kind of avenue typically prevails only to the extent that it is successfully championed by vested interests with an upper hand in the circle of policymakers. Given that it prevails, it forces agents of change to retreat into the shell of corporatism instead of supporting a locomotive of experimentation and competition between alternative solutions. Only in that way could they avoid to be regarded as strangers held at bay by corporatist policy.

Such a situation would increase the temptation to emigrate and offer limited incentive to move in the other direction. Brain drain would prevail over brain circulation. Indeed, a brain drain trend is firmly entrenched in European corporatist society, whose hyperregulatory bureaucrats prefer status-quo security to liberty. These officials are thus the fiercest opponents of the influence that free agents of knowledge innovation can exert on society.

Insofar as the fundamental essence of the corporatist society is that of undermining the legitimacy of the individuals both as citizens in the democracy (Saul 1995) and free agents of innovation in the economy, the control of knowledge by using corporatist policy instruments becomes self-righteousness.

Should this doctrine be called into question by its poor achievements, the boundaries of policy and the attitudes of policymakers would change markedly and the government role be restricted to the definition of a policy context conducive to the grassroots development of free agents of knowledge innovation.

Corporatist Policy vs. Growth-Promoting Reforms

Corporatism has refrained from essential domestic reforms that, by unleashing the power of knowledge innovation, would unlock much-needed growth. Reforms are rejected by a strong message of inevitability – a "shigata ga nai" (nothing can we do about it) attitude.

The top priority is to redistribute the benefits and costs of corporatist policies, rather than promote reform-supporting policies. Corporatist group members think inside the group's box instead of outside the box in order to build on the opportunities that the knowledge economy opens up.

Within corporatist groups, corporatist-functional interests discourage colleagues in the same profession or activity from forming innovative communities. It does not matter whether you are good or bad as long as you remain in the ranks.

Take healthcare reform. In a corporatist context, framework agreements with the doctors' association rule out the option of allowing health assurance funds to draw up contracts directly with doctors, which would be a move toward a free-agent knowledge economy in the healthcare industry.

Hence, the corporatist structure, with strong power conferred on special-interest blocs, does not appear a successful way of organizing society in the knowledge economy to encourage growth. Indeed, corporatist states are those that show a poor performance on production, employment, and productivity.

In the context of the connection between the regulatory environment, levels of information and communications technologies (ICT) investment and productivity growth, a *OECD 2003* report noted that a corporatist-induced heavy regulatory environment results in low growth potential:

"Countries with high levels of product and labour market regulation have had lower shares of investment in ICT in recent years than countries where the regulatory environment is lighter. Productivity growth in countries such as the US, Australia and Canada with a record of strong investment in ICT has remained robust despite the recent economic slowdown. By contrast, productivity growth in some other countries has weakened, in spite of investment in ICT. To make effective

Exhibit 57: Thinking Inside the Box

"Corporatism is an ideology which claims rationality as its central quality. The overall effects on the individual are passivity and conformity in those areas which matter and non-conformism in those which don't" (Saul 1995).

"The core assumption is that the co-ordinates of direction – North/South, Profitable/Unprofitable, Right/Wrong, Progress/Regress – remained constant like Latitude and Longitude on the map. Whereas the goal of knowledge, which is conceived as multicultural and multidisciplinary, is to transform people and organisations from within" (Voice of the Entovation 100 2002).

use of ICT investment, companies need to be able to innovate and adjust their organisational structures and workforces to new working methods."

In one respect, the poor score of those countries where corporate bodies shape the regulatory environment can be viewed as a consequence of extensive constraints on the manner in which companies, knowledge innovation agents, and consumers can pursue their goals. The next paragraph develops this argument with reference to subsidy-based negotiated corporatism.

Subsidy-Based Negotiated Corporatism vs. Open-Ended Market Guidance

The idea that knowledge innovation needs to be supported by the robustness of government, which is opposed to the fragility of markets, has created a high-powered arsenal of government subsidies distributing a variety of public handouts.

A subsidy machine, which is a systematic way for government to distribute favors to special-interest groups, subdues most of the knowledge markets. Simply put, it means the politicization of these markets.

Negotiated corporatism, one of corporatism's many variants (Formica 2003c: 247–249), is entrenched at the top of the subsidy mountain. This in turn produces a powerful me-too attitude among companies as they compete in the "partisan" politicking market for resource and support rather than in the free market for new commercial outlets and clients.

Subsidy-based negotiated corporatism and the underlying protectionism shield companies from competition-driven innovation. Fiat is a case in point. The Italian carmaker was the last producer to install catalytic converters in its cars and did not offer standard 2-year product guarantees until 2003 – well after most of its competitors.

In principle, government policy that subsidizes high development costs of knowledge innovation can be regarded as appropriate as long as there is a defined plan to phase the subsidies out. In practice, however, this outcome is problematic because that policy is shaped on the anvil of corporatist power, which portends a loss of companies' autonomy. Companies do not have the autonomy to employ knowledge innovation agents. The politicization of the knowledge markets connotes the management of subsidies by corporatist state-biased consultants engaged in subsidy-seeking deals on behalf of special-interest blocs. These "subsidy warriors" put knowledge agents pledged to create innovative ventures and governed by comparatively open-ended market guidance on the back foot.

Besides these worries, there is also the preoccupation that business subsidies in a corporatist regime of tax complexity, characterized by high social security taxes, high marginal rates of income tax, and high taxes on property and capital, drive a large number of the operating decisions, and hence innovation strategies. The more complex the tax regulations, the more potential loopholes – and the more time is wasted finding and exploiting loopholes instead of being focused on the capacity to innovate.

Exhibit 58: Negotiated Corporatism

Negotiated corporatism is that variant of corporatism for which concerted actions are shared between politicians, public officials, former government and party functionaries, and the most activist vested interests.

Negotiated corporatism is the product of coordinate market economies.

The price is very high for our societies and economies because innovation remains imprisoned in people's heads, never to see the light of day.

If there was a growing recognition among policymakers of the power of knowledge innovation as a fundamental driver of strategic change, governments would strive to downplay subsidies as a way of promoting knowledge innovation.

In the context of open-ended market guidance, a new knowledge policy would be structured putting greater emphasis on a greater role for the array of initiatives that free agents could implement within national frontiers and across borders.

Top–Down Corporatist Groups vs. Bottom–Up Communities of Free Agents

Policy interventions based on corporatist solutions are a conspiracy of the representatives of professional corporations against the laity of free agents of innovation, who cannot compete with them on even terms in the knowledge markets.

Among the most corporatist societies, Japan and Italy are cases in point. In these countries politically overrepresented interest groups comprising incumbent businesses, professional societies, and trade unions fear major changes in established relationships and thereby exploit their current relationships with bureaucracy in order to keep fees and entry barriers high and to exclude new entrants and outsiders altogether. Bureaucrats, meanwhile, maintain their power by assuring that those syndicates and regulatory structures reinforce each other.

Powerless against the corporatist state and hence reduced to the role of secondary participants, the knowledge innovation agents begin to decline. They have no alternative but to make the hard choice to get into the upper ranks of the professional corporations where they could have access to the corporatist group–state relations and perhaps gain some influence on the decision-making processes.

However, in taking this shelter, free agents would transform from bottom–up into top–down corporatist groups. They would be poised to confine their role within the stringent rules and procedures of the professional corporations. Coercive contracts would force them to do what the corporations want.

At this point, it would be proving impossible to form communities of free agents as vehicles of knowledge innovation ensuing from a long-lasting and spontaneous phenomenon of informal, cross-functional, and cross-boundary interaction between individuals.

Dissociated Corporatist Consensus vs. Unitary Community Consensus

The received wisdom of the corporatist consensus is that the efficient operations of markets – be they product or knowledge markets – require policy measures founded and built on consensus established by the corporatist groups. In the past, this argument has lent decisive support to the policymakers stepping into the product markets. Today, the apologists of the corporatist consensus argue that it would not have to be rethought in the case of knowledge markets.

This form of consensus is the outcome of an incestuous relationship perpetuated by mainstream politicians and representatives of corporatist institutions. Relying on instruments of co-optation instead of pointing to the need of effectual representation, they become closer to each other than politicians are to the public. Moreover, they work together to keep out new forms of association, as is the case for communities of free agents. These makers of knowledge innovation, who never gain a meaningful voice, are barred from promoting innovative practices that may have much to contribute to a regime change: from a closed-innovation regime of the corporatist associations rooted in industrial society to an open-innovation regime of the knowledge innovation communities.

Exhibit 59: Closed- and Open-Innovation Regimes

At company level, "closed innovation" is the notion that "successful innovation requires control from start to finish. According to this view, companies must generate their own ideas, develop them, build them, market them, finance them and support them on their own" (Chesbrough 2003b).

Closed innovation is the underlying logic of the corporatist model. A collaborative approach between different professional corporations that will lead to knowledge innovation is too alien to their "closed shop" mentality and restrictive practices whereby each interest group finds itself in contention with all others while protecting its existing entitlements. "Don't hurt me and I won't hurt you" is the approach that applies here.

Set against this view, "open innovation" assumes that "businesses can and should use ideas from outside [for] useful knowledge today is widely distributed" (Chesbrough 2003b).

As a McKinsey report argues: "Companies are opening their product development processes so that customers and other partners can offer their ideas and input. Some companies are now moving beyond the basic concept of 'open innovation' to an even more open process known as 'distributed co-creation'... The 'threadless approach' is instructive. Anyone can develop his or her own t-shirt design. Website visitors vote on their preferred designs, and the top vote getters move into production. Other firms, like Wikipedia and Red Hat, also employ variants of this approach" (Bughin et al. 2008).

Besides avoiding anything new or controversial, the incestuous consensus shows a dissociated nature. All special-interest groups have the political power to lobby for special protected benefits. The "give me more" attitude of each group is stifled because it relies on an unending process of mediation by which no single interest group should obtain any advantage over another.

The complementary effect of an unrelenting mediation is that new energies are not available to be channeled toward innovation. Corporatist leaders' ability and willpower are constantly addressed at deploying and directing resources toward the familiar routines.

The outcome is that too often policymakers become embroiled in a fierce brawl over "drilling holes in the ground as long as there [is] unanimous consensus" (Saul 1995). The corporatist society therefore faces a slow death by consensus-based decision making that has hit a wall.

Unlike the incestuous, dissociated, and institutionalized corporatist consensus, the informal consensus is unitary (given that it is reached in a context of spontaneous interactions between free and heterogeneous communities of knowledge practitioners). In fact, this form of consensus encourages discordant voices and collision of ideas that make the consensus area look broader and deeper. The breadth of consensus adds immeasurably to its quality, which can be channeled into the different communities in accordance to their own capabilities.

Standardization vs. Creative Ideas

Adam Smith was a trailblazer in the field of the distinctive arrangements that should underpin the liberal professions. His idea was that "the knowledge of liberal professions cannot be standardised in the same way as that of a mechanic trade" and that "the real controls on quality [of the professional services] are the disciplines of the market" (quoted from Dingwall 1996: 2).

Contrary to Smith's perspective, a conventional argument for professional corporations is that they exercise an effective control on the quality of the services provided by their associates. This control is made possible by an explicit body of rules through which the knowledge of liberal professions is somewhat standardized.

Hence, policymakers encourage competitive benchmarking exercises which, in turn, favor performance measures focused on the actions of competitors rather than on what the buyers want. This reinforces the corporatist behavior that spontaneously drives each and every member of a corporatist group first to rate herself/himself against the other members of the group and then reacting according to a "me-too" attitude at the expense of buyers' requirements of the buyers.

Shifting the focal point on the latter means that the discipline of the market substitutes for an introverted control. When hit with this change, those knowledge workers who previously were constrained by conventional corporatist rules could be remarkably adept at creating new market spaces by means of knowledge innovation.

But if the former attitude was to become even more attractive, thanks to interventionist policies, standardized solutions (identical for all market players) would come to be accepted as an established fact.

Knowledge Stock vs. Knowledge Flow

Knowledge providers in corporatist groups treat knowledge like physical capital: as a stock term. Since each guild typically relies on a specific type of codified knowledge, the knowledge stock is equal to the stock of individual pieces of that sort. Their decisions, which concern the desired level of the knowledge stock, are twofold:

- Reuse of existing knowledge:

  On the one hand, those providers decide to bring individual pieces of knowledge built up during the previous period over the next one.

- Optimal adjustment of knowledge stock:

  On the other hand, they decide how fast to build up new pieces of knowledge, how much to spend on it, and which adjustment pattern to follow.

By contrast, the free-agent perspective conceives of knowledge as a flow of interrelated and interdependent components. The desired rate of knowledge flow is the main decision and follows two lines:

- Complementarities among knowledge components:

  As said earlier, free agents contribute to the creation, evolution, exchange, and application of new ideas into marketable goods and services. Their mission is to put knowledge into action (they are knowledge innovation makers) by means of exchange. Hence, free agents allocate their organizational resources so as to facilitate exchanges of individual pieces of knowledge among themselves and also the different knowledge innovation communities to which they belong.
  The purpose is to get done what would be impossible for each of them (agent or community) to do otherwise: that is, devise a bundle of knowledge components whose value exceeds individual pieces of knowledge.

- Optimal behavior:

  Free agents are concerned as to what is the optimal behavior that leads to the desired rate of knowledge flow. Unlike rule-bound guild members, who tend toward enclosed doctrine, knowledge innovation agents are open to fresh and tacit or intuitive knowledge.
  Each agent acts with the consent of all the others affected by his conduct for creating, maintaining, and applying bundles of knowledge components that help knowledge consumers to make critical decisions concerning the creation of new markets, so as to extend their actions well beyond the incremental refinements of established products and services.

Policymakers of coordinate market economies emphasize government-sponsored economic programs that stem from tripartite arrangements, in which the public and private sectors are called in to coordinate their efforts for the purpose of developing new technologies and industries.

According to this policy view, government financial assistance is granted to government-run and government-backed agencies by direct subsidies and long-term implicit contracts with the government, not infrequently masquerading as public tenders. In both cases, professional corporations exert a great influence on and are instrumental in the realization of the agencies' projects.

Exhibit 60 – The bureaucratic facilitating agency

This is an intermediate body between the market and the state formed by collaboration between the government and the private sector organisations (the so-called public-private partnership). The agency is a syndicalist or corporative body that replaces the impersonal and anonymous mechanism of the market by its own deliberate actions in order to reach common chosen goals.

The agency is the result of government intention according to "the highly debatable idea that there are 'blueprints' (such as clusters) that can be readily implemented in quite different local economic, social and institutional contexts" (Martin and Sunley, 2002: 42). Consciously constructed blueprints are challenged by a mode of governance driven by a large number of agents spontaneously self-organised, automatic co-ordinated, interacting and evolving dialectally in the marketplace (we have called them "free agents") – as it happens in a complex adaptive systems (Stacey, 1996: 347).

Firms can succeed if they have access, apart from physical infrastructures, to intangible resources – basically, services such as specialised training, research, technology transfer, technical support, financial advice, marketing and design services, and information. These can be envisaged either as "private" services the provision of which is the outcome of seminal, market-driven interventions by profit-seeking free agents in the marketplace or as "club" services whose conceptual foundations can be detected in the theory of clubs (Buchanan, 1965; Casas-Pardo et al., 2001). Under the provision of government-sponsored economic programmes, clubs take the shape of agency-originated business service centres which support the costs of producing services partly through admission fees and charges for utilising a specific service paid by its members, and partly through a government fundraising mechanism. In the club framework, free agents are bound to the ruling elite of the agency and "free" only in terms of adjusting their activities to the agency's deliberate control.

The agency monopolistic stance inevitably develops a full range of pathologies such as patronage systems to prevent the emergence of competition, organisational dystrophy triggered by a bloated bureaucracy, and unresponsiveness. These diseases unavoidably sap free agents' energies.

(continued)

Exhibit 60 (continued)

The agency also distorts competition by managing subsidies and raising artificial barriers to entry into the "subsidies market". In one way, companies become addicted to handouts. In another, grants encourage the agency to lobby different government levels for more money, while, in consequence, a larger bureaucracy is created to handle those subsidies.

This unfolding array of arguments makes all the more urgent the need to highlight the profound cultural clash between the genuine entrepreneurial nature of the knowledge economy and the mode of governance through the agency's bureaucratic apparatus.

Convoluted and inflexible criteria and procedures encapsulated in the bureaucratic codes of those arrangements do not allow for timely decisions and decisive actions. This is the context that spotlights "knowledge stock" type of decisions, which will prevail over "knowledge flow" decisions.

Worst of all, subsidies and implicit contracts with the government set in motion a cumbersome process of intermediation by the agency. This deflects the decisions made by the agency's knowledge providers – the professional corporations – from the discipline of the market, because the subsidized agency shelters them from this rigor. This explains why a country that pursues such a policy is condemned to remain an underachiever in the knowledge economy.

Liberal market economies have traditionally taken a suspicious view of direct government involvement in the economic arena. Their policymakers show a preference for one kind of provider (the free agent) over another (the agency and its acolytes), and for acknowledging knowledge consumers over knowledge providers.

Market incentives, basically in the form of tax credits, are harnessed in a way that, for one thing, will benefit consumers and, for another, stimulate small-scale market arrangements implemented by free agents in the context of open-ended market guidance.

Knowledge decisions are very much a matter of "flow," and free agents must learn to "go with the flow" if they want to perceive the direction of markets, competition, and technologies that fundamentally change the way business is conducted in the markets they serve.

Knowledge Recycling and Diversion vs. Knowledge Creation

In the corporatist environment of consensually managed economies, a plethora of conflicts of interest exacerbates matters. Policymakers are more concerned with pleasing corporatist groups, and the latter with blessing elected politicians and their officials, than with creating room for knowledge innovation dynamism.

This results in uniform standards through the setting of rules that reduce the value of knowledge flow and its pluralist process of diversity, investigation, and experimentation. It closes the door to the process of knowledge creation. Policymakers take all the decisions that might facilitate the alternative process of knowledge recycling and diversion.

Knowledge creation means that a free-agent community lays the foundations for the fastest expanding industries of the future that would not have existed otherwise. In such an economy, diversity rather than specialization holds the key to long-run success.

Knowledge recycling and diversion means that a community represented by the primacy of special-interest groups espouses the causes of even inefficient but stalwart members (be they producers, traders, liberal professionals, or consultants). In such an economy, specialization within today's framework of industries and guilds will prevail over diversity. As a result, the community welfare improves at a snail's pace (at best) compared with the free-agent economy. At worst, it seems destined for an irreversible decline.

Even more disturbing is the influence of all sorts of vested interests on the fixations of theories and practices of entrepreneurship, innovation, and management, which prevent the advance of knowledge beyond the borders of the competencies required to make successful adjustments within the limits marked by the given context of current specialization.

Working otherwise, unconstrained by the predictable standards of business behavior fixed by the corporatist language – so as to open the door to knowledge creation – encounters dogged opponents in the interest groups. In such cases free agents cannot reach the critical mass needed for knowledge innovation, and their attempts to fill the gap between recycled knowledge and new knowledge look like an uphill task. That is why in the managed economies, innovative undertakings are extremely onerous and thereby few and far between.

International Knowledge Policy: Corporatist Partnership vs. Cooperative Partnership

In most recent times, the international community, including multilateral institutions, has drawn up guidelines in the hope of strengthening corporatist experiments in the international business arena. Instrumental in setting the trend for extending corporatism well beyond the nation-state have been, on the one hand, the European Union funding mechanism for programs and projects across EU countries, and on the other hand the UN Vision Project on Global Public Policy Networks. In both cases, EU supranational corporatism and UN global corporatism (Ottaway 2001) have been acknowledged as the best policy frameworks for closing the EU and UN governance gaps. Unfashionable corporatism has been renamed "public–private partnership."

As yet, public–private partnership policy has not tended to reflect alternative policy views. We cannot thereby lose sight of the fact that the innocent expression "public–private partnership" still turns out to be a corporatist arrangement of which government-backed agencies – a decisive instrument of political power in the corporatist states – are the most typical example.

In the international community, corporatist partnerships boost the growth of the special-interest groups whose parochial interests used to be confined within national borders. Moreover, they spotlight a new type of vested interests: those represented by non-governmental organizations. The overriding concern is that the static society that corporatist groups impose at the national level might be reproduced by corporatist partnership at international level.

As restrictive trade practices between the two world wars fuelled the collapse of trade, adding to the plight of industry, so today supranational and global forms of corporatism pose a growing threat to knowledge exchanges. There is a high level of risk that the knowledge economy may go off its expansionary trend due to restrictive practices concocted by corporatist partnership agreements, with a view to exercising the most effective sway over the great trade routes of knowledge. Vested interests that corporatist partnerships must accommodate impose a burden on the international circuit of the knowledge economy.

International knowledge policy that favors supranational and global corporatism sets the stage for international institutions that can be likened to the Paterson's authoritarian "Society of Status" where "the individual is not recognised [and] a man is defined by his relation to the group, and is presumed to exist only by permission" (Paterson 2003: 42). Corporatist privileges and subjection on an international scale, one would have to say.

Cooperative partnership shows a very different nature. This form of collaborative individualism, which embraces the communities of knowledge innovation agents, is founded on the intelligence of individuals from diverse backgrounds, their interactions in socially heterogeneous groups, and concentration on complying with the tasks required to structure the process of knowledge innovation, rather than on buying influence at the political market.

In this respect, a workable knowledge policy framework is that which suits the requirements of a liberal "society of contract," where knowledge can flow most fluidly through free contracts. Thanks to the abolition of corporatist control over the international circuit of knowledge, those agents can mobilize knowledge in response to market signals not distorted by overregulation and anti-free market practices. Hence, the stream of knowledge flows unfettered from continent to continent.

International and multilateral institutions cannot evade a decision that is bound to determine the course of the world economy for years to come. In as much as the international community were to become committed to sustaining the emerging trend and accounting for the challenges of the knowledge economy, policymakers would be going to consider the principles of freedom, openness, and institutional fluidity of cooperative partnership, as opposed to restriction, closeness, and institutional thickness of supranational and global corporatism.

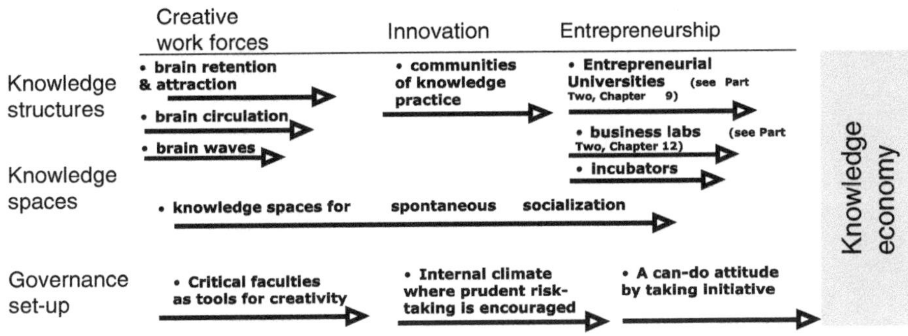

Fig. 11 A road map to the knowledge economy

A Road Map to the Knowledge Economy

Broadening and deepening the knowledge-intensive base of the economy, upon which rests the prosperity of knowledge-driven entrepreneurs (see Part II), is the policymakers' ultimate destination. A road map, tailored to the specific issues confronting a knowledge-oriented economy, is an instrument suitable for framing and, subsequently, tracking explicit policy actions.

The road map consists of a "compass rose" to identify crucially complementary progresses in the areas of intangible assets. Main categories of the compass rose are as follows:

- Creative work forces, innovation, and entrepreneurship
- Knowledge structures, knowledge spaces, and governance setup.

Within a matrix format of those categories, illustrated in Fig. 11, the arrows (resembling the compass needle's function) correspond to would-be policy makers' actions that help display the evolving process of a knowledge-oriented economy. Materialization of effective knowledge communities and exchanges requires the negotiation of (or even breaking with) counterproductive barriers between societal, national, and cultural spheres, so as to enable the spontaneous socialization and cross-fertilization that can help bring about and implement unexpected combinations of knowledge.

Note

[1] Primo de Rivera's dictatorship in Spain (1923–1929) and Mussolini's Fascist Italy (October 1922 to July 1943). Mussolini defined corporatism as the merger of government and corporations.

Chapter 8
Global Advance of the Knowledge Economy

The great challenge for the next several decades will be to advance understanding of social systems in the same way that the past century has advanced understanding of the physical world.

(Jay W. Forrester)

As has been discussed already, the impact of the knowledge-based era is pervasive. It does not apply only to individual sectors or only developed countries. Although there is a massive difference in the readiness and ability of different actors to take advantage of the new opportunities and counter the difficulties, change has arrived across the board.

Only a few years ago, it was correctly said, "half humanity has still not made a phone call." That, being true then, is certainly not true anymore. In great parts of the world, there has been a deep concern about the "digital divide," reflecting the perception among billions of people around the world that technology has not helped them catch up with developed societies and the anticipation that the latest technological (information) revolution would not be different. True, ICT has in some cases been put to use to acquire computers or develop data banks that were not really needed. There have certainly been excessive costs and inappropriate training, which has left a lot of expensive investment idle and doomed to decay. Such problems are far from done and over with. The fact is, however, that the outlook has shifted in decisive fashion.

With the accelerating diffusion of mobile telephony, the information society is in the process of reaching out. Competition has set in; costs of hardware are coming down; services are becoming user-friendly, even for the illiterate, and with consideration to local conditions calling for energy-intensive and environmental-friendly technologies. People anywhere are getting access to information they could not obtain in the past – information that in many cases is of great value to them, although there could without any doubt be a lot more production of such useful information and a lot less of information that is not useful.

T. Andersson et al., *Knowledge-Driven Entrepreneurship*, Innovation, Technology, and Knowledge Management 102, DOI 10.1007/978-1-4419-1188-9_8,

In this chapter, we examine the state of readiness for different parts of the world. Special attention is devoted to the degree to which each of the major regions of what is now referred to as the developing world have advanced in important respects.

Table 1 summarizes some key characteristics for the following developing country regions: East Asia and the Pacific, Europe and Central Asia, Latin America, Middle East and North Africa, South Asia, and Sub-Saharan Africa, and for China and India separately.[1]

Generally, we describe the situation up to 2006, the latest years for which data are consistently available. This information highlights the important differences across regions in terms of income level, growth, productivity,[2] economic structure, and human development and poverty. As seen, East and South Asia have experienced higher growth than other regions. Together with Eastern and Central Europe, they have also displayed higher labor productivity growth. Regions differ remarkably in the scores on UNDP's Human Development and Poverty indices.

Key Features

Aspects of the knowledge-based economy span a wide range of information, including education, conditions for enterprise creation, finance, research, and more. By necessity, an overview of KBE (knowledge-based economies) data draws on several databases. The World Bank's so-called Knowledge Assessment Methodology (KAM) database integrates many of these sources, including: World Development Indicators and Governance Indicators (World Bank), the Human Development Report (UNDP), The Global Competitiveness Report (World Economic Forum), research and education data from UNESCO, ICT data from ITU, among other things. In bringing together these data sources, we get some basis for international benchmarking of KBE, including with regard to strengths and weaknesses of individual countries.[3] The KAM database is organized along four main dimensions relating to the knowledge economy:

- An *Economic and Institutional Regime* that provides the general incentive framework for new ideas to be born, used, and/or adapted, including, for example, investment levels, trade policy, government effectiveness, democratic institutions and procedures, the level of corruption, etc.
- An *Education System* that upgrades the essential human capital input – a skilled and educated population – that can access, use, and share knowledge, invent new products and services, or adapt foreign technology and ideas to local conditions. This includes, for example, literacy rates among the adult population, enrolment rates, and the availability and quality of different kinds of skills.
- The infrastructure for, and use of, *Information and Communications Technology,* especially the use of telephones and computers, but also exposure to media through which ideas are transmitted.
- An *Innovation System,* consisting of firms, academia, government and other actors, that impacts on whether innovations thrive. Indicators of the innovation system include, among other things, FDI intensity (as a potential source of tech-

Table 1 Regional development indicators

	GDP/Capita in USD	GDP rowth (%)	Labor productivity	Labor productivity growth	Human Development Index
	2007	2000–2007	2006 2/	2000–2006	2008 3/
China	7,660	10.0	0.7	8.9	0.78
India	3,827	7.2	0.5	5.4	0.62
Europe and Central Asia	10,769	5.9	2.7	5.0	0.81
Latin America and Caribbean	7,685	3.5	2.5	0.9	0.80
Middle East and North Africa	11,476	4.3	1.4	0.6	0.70
East Asia and Pacific	16,493	8.7	0.7	7.2	0.77
South Asia	3,055	6.8	0.4	4.6	0.61
Sub-Saharan Africa	3,223	4.8	0.5	2.0	0.49

Source: World Bank (2008)

nology) and R&D intensity (spending on research) in the economy, science and engineering graduates, scientific publications, and the share of high-technology products in exports.[4]

Table 2 summarizes the KBE data for different regions, including China and India as special but highly dominant examples of East Asian and South Asian economies (over three quarters of GDP and population in each region). A closer study of the table reveals that most indicators worsen as average income falls, with East Asia generally scoring highest and Africa lowest. Richer countries/regions have, among other things, been able to expand education and ICT infrastructure. Broadly speaking, countries with higher education and better conditions for entrepreneurship and innovation have also seen higher long-run growth than countries where such conditions are not in place.

Again, it should be emphasized that the quantitative aggregates convey limited information, since quality differences are known to be substantive without being captured appropriately in the statistics. It should also be underlined that the direction of causality between different variables requires careful consideration. Yet, while the challenges facing South Asia differ from those of, for example, East Asia, each region has areas where there is room for improvement. An attempt to depict stylized regional profiles follows later, with consideration to the four dimensions outlined earlier.

East Asia and the Pacific

In this comparative overview, East Asia and the Pacific (EPA) is the "star-performer" region compared with others and the developing region that can be used as an international benchmark in most areas.

The region has made a strong performance in both economic and social welfare terms as seen in Table 2, and the general incentive structure for knowledge creation and use is also favorable. Overall, the economic and institutional regime is relatively strong. EPA countries have, on average, the highest levels of investment and openness, liberal trade regimes, competitive environments, and some intellectual property rights protection in place. Governments are comparatively effective in formulating and implementing policies. Political freedom and democratic institutions are ranked beneath those of other regions at comparable income levels, however.

The EPA region has also come a long way in terms of education, although it continues to lag behind countries in Eastern and Central Europe (ECA). Adult literacy rates (those aged more than 15 years) are at 90%, and youth literacy rates (for ages 15–24) are nearly universal.[5] Gross enrolment rates for secondary and tertiary levels of education are high by comparison with other developing regions.[6] The quality of education, to the extent it can be measured by achievements in maths and science, is also comparatively high. Yet, it is important to note that only two fifths of the young people in age of studying at university are actually pursuing tertiary education – the equivalent average for high-income OECD countries is 60%. Another aspect is that the effectiveness of East Asian educational systems is questioned in some respects – "vertical learning" and a preoccupation with passing tests are widely thought to hamper genuinely explorative and innovative capabilities.

There are more telephones than elsewhere, twice as many computers as in the next region (ECA-Easter Europe and Central Asia) and many more Internet users. These strengths extend into the specific conditions for innovation. Foreign direct investment is high, albeit lower than in ECA, as it is the share of high-technology exports and the absolute and relative availability of researchers. The expenditure on R&D is also higher, including in the private sector, and there are four times as many patent applications emanating from the EPA region as from the other regions together. The academic output from EPA as measured in publications by far exceeds that of other regions.

East Asia has a greater proportion of knowledge-intensive industries than most other regions of developing countries. A particular feature worth noting is that ICT and associated innovations are put to active use in productive activities to a high degree. ICT is put to concrete and direct use as tools to enhance productivity, build new channels to markets, gain access to new technologies, and not merely as an instrument for diffusion of information in a general sense (Joseph 2002). This also reflects that East Asian countries are generally stronger in applied research and incremental innovation than in basic research.

Within East Asia, China deserves special mentioning. This is not only because China accounts for a high share of the population in the region, but also because of its special characteristics (related to size, the presence of a large "modern" sector in absolute terms) and also the outstanding economic performance displayed by the country in recent years. Its average income level is still today only around half of the regional average, however, and its economic incentive framework is also less favorable than that of East Asian countries in general, notably as China is considerably less open. Education levels and ICT infrastructure are closer to the (high)

Table 2 Summary KBE data by region

		China	India	East Asia	MENA	ECA	LAC	Africa	South Asia
Economic Performance	GDP per capita (in current $ PPP), 2006	7,660	3,827	16,493	11,476	10,769	7,685	3,223	3,054.6
	Human development index, 2005	0.78	0.62	0.80	0.76	0.80	0.77	0.51	0.60
	Composite risk rating, 09/2006–2008/2007	79	70	76	74	71	70	65	64
	Unemployment rate (% of labor force), 2004	4.2	5.0	5.9	9.5	13.1	10.3	13.6	6.9
	Employment in services (%), 2005	16	20	47	57	49	60	39	31
Economic Regime	Gr. capital formation as % of GDP, 2002–2006	23	29	27	23	25	21	21	24
	Trade as % of GDP, 2006	70	44	151	97	110	85	78	52
	Tariff and nontariff barriers, 2008	70	51	77	70	80	74	68	49
	Intellectual property protection (1–7), 2007	3.4	4.0	4.4	4.2	3.3	3.2	3.2	3.1
	Domestic credit to private sector (% of GDP), 2006	114	46	78	51	37	37	23	36
	Days to start a business, 2008	35	33	37	32	26	54	46	40
Governance	Regulatory quality, 2006	–0.19	–0.15	0.35	–0.05	0.06	–0.05	–0.52	–0.42
	Rule of law, 2006	–0.40	0.17	0.28	0.06	–0.28	–0.43	–0.54	–0.44
	Government effectiveness, 2006	–0.01	–0.04	0.51	–0.06	–0.06	–0.17	–0.50	–0.52
	Voice and accountability, 2006	–1.66	0.35	–0.21	–0.81	–0.02	0.15	–0.40	–0.57
	Political stability, 2006	–0.37	–0.84	0.17	–0.45	–0.14	–0.26	–0.32	–1.65
	Control of corruption, 2006	–0.53	–0.21	0.23	0.09	–0.26	–0.28	–0.48	–0.69
Innovation	FDI inflows as % of GDP, 2000–2005	3.2	0.9	4.6	2.3	5.1	3.9	3.2	0.8
	Total royalty payments and receipts (US$/pop.), 2006	4	0	258	23	22	15	6	0
	Science and engineering enrolment ratio (%), 2006	n/a	20	28	24	22	24	20	22
	Researchers in R&D/Mil. people, 2006	715	n/a	1959	690	1557	310	108	93
	Total expenditure for R&D as % of GDP, 2006	1.3	0.6	1.0	1.3	0.7	0.3	0.4	0.5
	University–company research collaboration (1–7), 2007	4.1	3.5	4.2	3.2	3.1	2.8	2.7	2.9
	Scientific and technical journal articles/Mil. people, 2005	32	13	234	86	107	21	6	5
	Availability of venture capital (1–7), 2007	3.0	4.1	3.9	3.7	3.1	2.8	2.6	3.0
	Patents granted by USPTO/Mil. people, avg. 2002–2006	0.4	0.3	44.6	10.0	1.2	0.9	0.1	0.1
	High-tech exports as % of manuf. exports, 2005	31	5	31	4	8	7	7	2

(continued)

Table 2 (continued)

		China	India	East Asia	MENA	ECA	LAC	Africa	South Asia
Education	Adult literacy rate (% age 15 and above), 2005	n/a	61	93	80	98	87	59	60
	Gross secondary enrollment, 2006	76	54	82	79	91	81	38	51
	Gross tertiary enrollment, 2006	22	11	43	26	47	29	5	7
	Prof. and tech. workers as % of labor force, 2004	n/a	n/a	21	19	23	16	11	7
	8th Grade achievement in mathematics, 2003	n/a	n/a	513	411	495	387	302	n/a
	8th Grade achievement in science, 2003	n/a	n/a	510	430	497	413	288	n/a
	Quality of management schools (1–7), 2007	3.6	5.7	4.7	4.2	3.8	4.1	3.5	4.0
	Brain drain (1–7), 2007	3.9	3.8	4.1	4.0	2.9	3.4	2.6	2.8
Gender	Gender development index, 2005	0.78	0.60	0.80	0.74	0.80	0.76	0.50	0.58
	Females in labor force (% of total labor force), 2006	45	28	43	26	46	40	42	33
	Seats in parliament held by women (as % of total), 2005	20	9	19	8	15	18	18	13
ICT	Total telephones per 1,000 people, 2005	570	128	823	682	804	596	174	115
	Computers per 1,000 people, 2005	41	16	274	137	142	84	36	13
	Daily newspapers per 1,000 people, 2000	59	60	115	41	116	57	15	43
	Internet users per 1000 people, 2005	85	55	314	152	207	164	30	29
	Price basket for Internet (US$ per month), 2005	10	7	16	17	15	27	45	11
	Extent of business Internet use (1-7), 2006	3.5	4.6	4.5	3.6	3.7	3.7	3.1	3.7
	ICT expenditure as % of GDP, 2006	5.3	6.1	7.4	4.2	5.2	6.1	5.7	5.3

Source: World Bank (2008)
The table includes the majority (but not all) of the indicators present in the KAM database

regional average, but there is some distance: tertiary enrolment rates are only 22%, compared with a high 43% for EPA as a whole, and there are only four computers per 100 persons. At the same time, backed by sociological and cultural factors, there is a notion of high and rapidly rising quality in the educational effort.

China has a particularly strong standing in innovation within manufacturing. The advances made in creating favorable conditions in innovation are remarkable given the otherwise relatively "typical" lower middle income profile. This is most particularly evident in the high share of R&D in GDP, the extremely large pool of researchers in R&D activities (over 700,000), collaboration between universities and private companies in research, the strength of the science park and incubator program of China, and the high proficiency and quality of the academic community.[7] Underpinning this performance also lie a series of reforms to the enterprise sector and regulatory frameworks gradually opening up more industries and markets to open competition (OECD 2005). The still burdensome sector of state-owned enterprises and the public sector, more broadly, account for a shrinking part of the economy. As will be further discussed in the last chapter, inward FDI and foreign technologies are also making major contributions to China's technological development today.

Eastern Europe and Central Asia

The group of former socialist countries in Eastern Europe and Central Asia show several strengths, especially in education, but also in ICT infrastructure and use. Adult literacy rates are universal, and enrolment rates are the highest among developing countries – almost half of the relevant age group is in university. The quality of education is at par with that of East Asia. Compared to East Asia, ICT infrastructure is less developed, but exceeds that of other regions. There are 14 PCs for every 100 persons, and 20% of the population uses the Internet.

Whereas the education level in transition countries is relatively high despite low incomes, there is a lack of certain needed skills, which accounts for bottlenecks.[8] This also applies to ICT use. One reason is that, despite strengths in maths, engineering, and computer science skills, there is also a lack of life-long learning mentality. A big drag in the transition economies broadly is the weak links between the production sector and the public R&D sector and the flow of new profitable applications of technology. During the socialist era, all applications of R&D were of course controlled by the state, which relinquished this role only in the early 1990s and left it completely to the different institutes. They have not managed to diffuse technology, because lack of incentives, managerial competence, and often insufficient financial support.

ECA's relative weaknesses (compared to the EPA region, and compared to its strengths in educational achievements) lie in the economic and institutional setup and in the specific conditions for innovation. Investment rates are slightly lower than in East Asia, and the economies are in general less open to foreign and domestic competition. Perhaps most importantly, government effectiveness and enforcement capabilities are also lower and corruption represents a serious problem. In Central Asia and Eastern Europe this has been shown to be associated with bribery, uncertainty because of a lack of legally enforceable agreements, poorly functioning markets especially as regards public procurement, and lack of bureaucratic accountability and poor public services. These factors account for a particularly serious problem for new private firms and, with respect to firm size, constitute a significant regressive tax (Hellman et al. 2000). Further related to innovation, FDI intensity is less than half of that of East Asia and the high-tech share of exports on average less than one third. There are blatantly few patent applications to the USPTO (although if the European Patents Bureau was taken into account the difference may arguably be smaller). There is less venture capital available and less collaboration between universities and companies and between companies (clusters).

The group remains clearly marked by their history of stark public ownership and dominance over resource allocation, with property rights concentrated in the hands of a few central planners. Especially in the Asian former Soviet areas, the post-socialist institutional revolution is not over (Piatkowski 2002). High transaction costs continue to stifle the economy. Weak contract enforcements and lacking respect for property rights raise the costs of market exchange; less trade is performed and entrepreneurial efforts are shifted to less transparent gray and black markets.

Whereas there was heavy investment in basic science, education, and training, other kinds of skills were not developed. There is also the heritage of massive state subsidization of heavy industries, whereas many consumer products were sorely lacking. Because of remaining red tape and government intervention, but also prevailing attitudes, SMEs still suffer from severe lack of reliability in transactions and the underdeveloped supply of relevant professional services in market intelligence, logistics, technology absorption, and less diversified financial markets, which sharpen resource and liquidity constraints.

Transition economies have a marked history of adverse sentiments in regard to entrepreneurship, although it existed and sometimes even flourished under harsh circumstances (Kornai 1992). Today's entrepreneurs in part comprise the survivors of the old file and ranks, in part constitute the young whose practices contrast with the old ways. In both cases, they are still plagued by the consequences of weaknesses in governance, while also confronted with widespread suspicion in their societies. All this has serious implications for trust and networking and for the viability of top–down vs. bottom–up approaches (Smallbone and Welter 2001). The next few years present special opportunities as well as challenges for SMEs now engaged in the rapid opening of new markets through integration with the European Union.

Latin America and the Caribbean

Judging by the data, there is a gap between LAC on the one hand and ECA and East Asia on the other in terms of KBE conditions. Latin America's relative strengths appear to be in ICT intensity and, to a less extent, in educational achievements.

Thus, ICT expenditure is comparatively high, and there are six phones (fixed or mobile) and 0.8 personal computers for every ten people. Literacy rates and enrolment rates are considerably higher than in Africa and South Asia: literacy rates are at 87% and gross tertiary enrolment rates reach 29%; these are lower than in ECA and East Asia, however. Income differences are notoriously high and there are serious problems with societal cohesion, a large part of the population being employed in the informal sector, crime rates being high, etc.

Relative to other regions, Latin America is even worse off in terms of the overall economic incentive framework and the innovation system. Investment rates are low. The public sector is generally inefficient. Regulatory regimes are often weak and countries are on average less open to foreign or domestic competition than in the ECA or EPA regions, although trade regimes are relatively liberal. There is a relatively strong presence of foreign multinationals, but in many cases ventures were established under conditions of heavy protection, and their market dominance still remains high in many niches (Katz 2002). The quality of governance is a considerable problem, as is pervasive corruption. Governments in Latin America are more accountable to their constituents than in any other region, however.

In terms of the innovation system, LAC is on par with ECA in terms of the share of high tech in exports. The R&D intensity is much lower, however, as is reflected in very limited patent applications. Enrolment in higher education is higher than in poorer regions and scientific progress has been demonstrated in a number of countries, including Brazil, Argentina, Chile, Costa Rica, and Mexico. Not only do subjects such as legal affairs attain disproportionately high numbers, but the enrolment numbers for science and engineering are relatively high. However, the industrial relevance of the public R&D effort remains low. It is dominated by public funding of university research, which is not linked to the limited R&D effort pursued by the private sector.

There are pockets of impressive industrial advance in LAC. These include the science/technology parks in Campinas in Sao Paulo state, or the forestry sector broadly, in Brazil. Another example is food industry, which is highly competitive, but also finding ways to upgrade value-added services and renew itself, as in the case of the Chilean salmon industry. Vegetables in Guatemala, flowers in Colombia, fresh fruits in Chile, telecom services in Jamaica, and tourism in several countries, present other dynamic spots. Other new niches have evolved taking advantage of falling transport and communication costs and new opportunities for market access. El Salvador and Mexico have advanced labor-intensive hub processes, although these have come under pressure lately.

At the same time, further development of competitive agribusiness is held back by protectionism in developed markets. In recent years, labor-intensive industry has contracted, whereas capital-intensive resource-based industries account for little expansion, and none in terms of jobs.[9] While innovations of various sorts have been greatly important in all these cases, on the whole knowledge-intensive industry keeps expanding slowly. Linkages between universities and the private sector are weak, as are linkages between foreign investment and R&D, on the one hand, and domestic industry on the other hand.

Middle East and North Africa

The Middle East and North Africa region shares similarities with the Latin American region in terms of KBE characteristics. While the MENA region may appear more open based on trade and investment statistics, this is basically the result of investment in natural resources and oil exports. In reality, the MENA region is marked by difficult conditions for trade, investment, and business activity; more broadly, there is excessive public intervention, protection, and regulation, and public investment dominates massively in most sectors (Sala-i-Martin and Artadi 2002). In governance respects, including control of corruption, rule of law, and the effectiveness of public institutions, MENA is better off than the Latin American region, with a notable exception being the accountability of its governments and the stability it offers. MENA is also doing well relative to LAC in terms of ICT infrastructure, with a higher incidence of phones and computers. On the other hand, a lack of complementary regulatory reform in most countries keeps telecom costs on the high end – there is still some difference to East Asia and ECA, however.

MENA is somewhat less good at education, as evidenced in lower literacy rates – this is basically a gender issue, as male youth literacy rates are at par with more developed regions, while that of young girls remains much further behind. There is also a marked gap at tertiary level education compared to ECA and EPA. A lack of information channels, and also the presence of cultural and language barriers, contribute to liming ICT use and innovation. Beyond that, however, there is an overriding mode of authoritarian government as well as societal structures, which is at the same time overly protective in a number of respects. This serves both to limit individual independence from childhood, especially for girls, and to foster passive attitudes and lack of decision-making skills (UNDP 2003). Likewise, government bureaucracies are slow in applying new instruments and solutions in public services and in diffusing information (Aubert and Reiffers 2003). Historically, the region has pioneered in nearly every field of science. MENA countries have also expanded and established a substantial infrastructure for all types of S&T institutions, including R&D, over the last decades. Looking at the Arab States, R&D institutions increased from 26 in 1960 to 322 in 1996. Currently, more than 600 organizations are involved in R&D activity. The Arab states have established a considerable number of new universities, increasing from 10 in 1950 to 175 in 1995. The number of university colleges also grew from 29 in 1960 to 177 in 1996. All these units, centers, and councils were linked to each other on the macro level and also linked with national economic plans. In addition, intersectoral research units, laboratories, and research services were set up within academic institutions.

Despite this expansion, conditions for innovation are, broadly speaking, on par with those of LAC and, as a result, much behind those of ECA and EPA. FDI is smaller and the share of high-tech exports is lower. MENA has some relative strengths in areas such as publications, patent applications, and the pool of researchers in the region. There are strong countervailing barriers, however. The educational system appears to be largely failing both with respect to basic training and higher studies. Universities are perceived by industry to form ivory towers, whereas the private sector itself neglects training. Linkages between foreign and domestic industry are likewise very weak. Linkages are equally weak in the area of R&D. There are also barriers to widespread entrepreneurship, especially for women. Employment numbers are generally contracting and unemployment is high and on the rise, including for the highly educated, especially women. Today, MENA suffers from fundamental physical deficiencies, such as water shortages due to the inability to construct distribution systems, along with institutional ones which raise barriers to creativity, responsibility, and learning processes.[10] Regional differences are huge. Hubs such as Dubai, Cairo, and Casablanca have attained a massive lead in the adoption of new technologies as well as in generating employment, whereas vast regions experience limited or no progress in such respects (Accascina 2005).

There are important income and outcome gaps between the more developed regions reviewed earlier, and South Asia and Africa. Average income levels are less than half of those of LAC and one third of those of East Asia. These differences are mirrored in less favorable outcomes in terms of KBE economy achievements. Again, there are regional specificities, notably in the relatively strong performance of South Asia in innovation.

South Asia

Compared to richer regions, innovators in South Asia are faced with a weaker economic regime and institutions, much less availability of information and communications technology, and low education levels. Countries are generally less open, and trade regimes are more restrictive; fiscal deficits are higher, and bad governance is a serious problem. The ICT infrastructure lags behind the other regions, even including Africa. There are only 12 phones and 1 personal computer per 100 persons. Three percent of the population presently uses the Internet.

Educational achievements are considerably lower than in richer regions. Only 60% of the adult population can read and write, although for young people, literacy rates have increased. Those enrolled in secondary education make up only half of all children of relevant age, and only 7% are in university. These numbers mask, however, a rapid increase in high-quality learning in narrow segments, which still constitute large absolute numbers.

South Asia's countries have also considerable strengths in innovation. Even a country such as Bangladesh, despite its deep poverty, informal sector reliance, and strong cultural barriers to new ideas and, notably, disadvantages for women in taking commercial initiative, is developing sprawling entrepreneurship and small business in the wake of gradual market openings and improved supply of microcredit. A notably strong relative position is recorded for India. Similar to the dominant position of China in East Asia, India makes up over 70% of the total population of South Asia. Relative to other South Asian countries, India is also performing better in most aspects of KBE. On the other hand, India cannot compare with China in any area, except governance. Secondary and tertiary enrolment rates are low, at 54 and 11%, respectively. Internet penetration is much below Africa. But India has a remarkable comparative strength in rapidly growing segments of education and innovation. As in the case of China, the efforts of creating an environment conducive to innovation are remarkable in comparison to the otherwise standard low-income country setting. India has a very large pool of researchers and an academic community that is linked to the private sector, coupled with progress in cluster development.

On the other hand, vast stretches of society are yet weakly affected. As in MENA, cultural and societal structures hamper initiatives in large parts of the population. In several of the countries in South Asia, women are severely restrained in education and business activity, although they assume a major responsibility for families and long-term savings and investment decisions.

Sub-Saharan Africa

Sub-Saharan Africa is the poorest region of the world and it has the largest proportion of its population living in poverty, or in LDCs. Not unexpectedly, Sub-Saharan Africa is the region where knowledge appears to be furthest behind in almost all areas. The economic incentive regime is worse than in other regions except South Asia, although not spectacularly so. African countries are on average more open than those of South Asia and even LAC. Governance is a main weakness, however. As in South Asia, corruption is a key problem, and the regulatory quality is low; governments are considered ineffective and laws are not followed. The state itself and its core institutions are in many cases an artifact set up in the days of colonization, representing or dominated by only parts of the population, depending on the way in which traditional tribal barriers coincide with the modern national borders.

Despite the weaknesses, Africa has recorded enhanced GDP growth in recent years, improved macroeconomic management, increased agricultural production, and the cessation of conflicts in several countries (Economic Commission for Africa 2002; UNCTAD 2005). African exports have grown markedly in recent years, especially to the USA but also to other regions. South Africa, the economic giant of the region, after years of struggle is now doing rather well, and its macroeconomic fundamentals seem robust. Others, such as Botswana, are doing exceptionally well. Still, most of the African economies keep struggling with getting the fundamental economy in place. High indebtedness, caused by loans accumulated in past days of heavy and costly import substitution schemes or military build-up, has for many years placed them in states of extreme dependency on multilateral institutions for keeping afloat. Now, following the debt settlements in September 2005, many of the most battered economies stand a chance of gaining better access to resources and flexibility in how to use them.

Information and communications technology has been introduced and exceeds that of South Asia – there are three times as many computers per capita – but remains at a much lower intensity than in richer regions. Education and the innovation system manifest particular weaknesses in most of Sub-Saharan Africa (Andersson and de Silva 2006). Three in five adults are literate but youth literacy rates are higher, partly due to a strong catch-up in female literacy. Yet, secondary enrolment rates are much lower (in fact lower than tertiary enrolment rates in East Asia or ECA), and people in university make up only 5% of the relevant age group. The quality of education is also behind other regions. Fundamental imbalances in access to education as well as ICT infrastructure prevail between urban and rural areas. Yet, in some southern countries, notably Namibia, Botswana, and Lesotho, literacy rates have increased impressively in recent years, especially for girls In SSA as a whole; however, some 100 million young women lack literacy skills (Pigato 2001).

As for innovation, the African system stands in contrast to the South Asian one. The low enrolment rates at tertiary level are compounded by a lack of focus on science and engineering. On venture capital, publications, and patents, Sub-Saharan Africa is at the bottom of the league. Links between universities and the private sector are very week. Firms generally rely on personal networks and face-to-face meetings. The information available on the Internet has so far been viewed as largely irrelevant to most of the business sector.

Sub-Sahara's acute economic problems have worsened by the costly consequences of urgent health problems. Some two million people, most in Sub-Sahara Africa, die from Malaria each year. Among the many others that plague the continent, pneumococcal disease – a bacterial variant of pneumonia – claims an even greater number of lives. So-called meningococcal meningitis causes epidemics from Senegal to Ethiopia, killing 5–10% of those infected, leaving many others deaf or in other ways disabled for life. Some 300,000 persons, many of them children, have been hit over the last 10 years. On top of this comes HIV, which takes out a growing proportion of the adult population in many countries and keeps spreading. In many of these cases, vaccines are readily available, but not at any price affordable for most Africans.

South Africa's integration into the global economy has made science and technology education a growing priority. The move from labor-intensive to knowledge-based production depends on technologically sophisticated production procedures, in agriculture as well as in industry. The South African government has recognized the importance of improving education standards and meeting the demand for skilled labor. New jobs are mainly created in sectors requiring specialized skills, such as the export and financial sectors, while jobs are disappearing in older sectors depending mainly on low- and semi-skilled intensive labor. Difficulty in finding workers with appropriate skills keeps acting as a major constraint to growth.

Finally, as already noted, the regional characterization presented in this chapter is, of course, stylized in nature. Several caveats must be kept in mind. Naturally, within each region, there is a considerable variation between countries in terms of, for example, incomes, structures, and in basically any kinds of issue influencing their approach to the development and use of knowledge. In the same way, however, national averages are bound to mask important variation of such kinds within countries. It should be underlined that the potential of the knowledge-based society is particularly difficult to capture in the poorest communities. At the same time, KBE opens up particularly stark new opportunities to precisely these communities, which is now becoming pronounced with the diffusion of cellular technology and the convergence of communication means underway. The accumulation and the diffusion of knowledge are now in the process of offering a way forward in response also to grave development issues.

On this basis, we argue that the knowledge-based society is now a global phenomenon, of relevance to all, including the billions of people who still live in countries and regions that are commonly referred to as "developing."

Notes

[1] Only low and middle-income countries (including LDCs) are included in the regional averages.

[2] For lack of employment data, labor productivity is calculated as total output over labor force. This means that countries where participation rates have increased over time come across as having seen higher productivity growth than is necessarily the case.

[3] This source provides a mix of quantitative data and qualitative data and relies on a wide variety of sources: World Bank data, including on governance, UNESCO, IMF, and data presented in the World Economic Forum Competitiveness data, among others. Some of these sources are not representative, however, such as the industry surveys undertaken for the WEF global competitiveness report. Moreover, regional averages are based on countries for which data are available, which needs to be kept in mind when results are generalized to regional level. See appendix on the KAM methodology.

[4] In what follows, some innovation data, such as patents and number of researchers, are not weighted by population. There are economies of scale in knowledge creation and use. For example, a minimum pool of university researchers (no matter how small the country) may be needed to insure that some specific competence develops and that there is some contact with institutions abroad.

[5] Youth literacy rates are from the World Development Indicators, 2008.

[6] Gross enrolment rates are all those enrolled at a certain level of education as share of total population in the official age group corresponding to that level of education.

[7] The Chinese science park program, which now entails 38,000 enterprises, is designed to spearhead market-led innovation through support of "hard" as well as "soft" infrastructure (Zhou 2005).

[8] In Russia, the availability of researchers is similar to that in the major OECD countries, but R&D expenditure is lower. Over the last two decades, the number of researchers has been on the decline.

[9] Some, such as de Ferranti et al. (2002), argue that LAC should have made much more conscientious efforts to develop its natural resource-based industries. In some cases, such as copper in Chile, ownership was conferred to foreign investors under conditions that meant hardly any obligations or even returns in the form of tax revenue, etc.

[10] Pedagogical methods appear obsolete and hindering awareness and creativity. Other problems are old curricula, the absence of periodicals, and limited involvement of private initiative in education. Moreover, training and vocational training has long been neglected by most MENA countries (Djeflat 2002).

Part II
Entrepreneurship

Chapter 9
The Entrepreneurial Revolution

> *Nothing works the first time. The first 15 years of companies and humans tend to be the same – filled with hope, measles, failures, mumps, reorganisations, scarlet fever, executive troubles.*
>
> (Georges Doriot)

Historically, the rise of great civilizations has been predicated on *entrepreneurship* – that is, the ability of transforming novel ideas into goods and services that through trade can be bought and sold.

The rise of the Phoenician civilization, as well as the unparalleled growth of the Ancient Maritime Republics (Amalfi, Genoa, Venice, and Pisa) and the city of Amsterdam, was all credited in large part to the entrepreneurial spirit, energy, and initiative of their populations. The Phoenicians are renowned for their seafaring skills and mercantile entrepreneurship. Bold explorers, they pursued entrepreneurial actions that helped break old ways of thinking and open the way for new ideas. The merchant based city-states of Italy thrived – thanks to their adroit entrepreneurship in manufacturing, shipping, and international trade.

The entrepreneurial economy shifts the emphasis from the mastery of physical assets (e.g., natural resources, infrastructures) and physical tools (e.g., computers) to that of intangible assets (such as education, research and development projects, brands, patents) and sociocultural tools (such as communities of knowledge practice) as the key to a community's economic prosperity.

The process of discovering and exploiting breakaway opportunities depends upon the production of new knowledge and how it brings about spillover effects. The method and forms by which a society organizes the production, dissemination, and transfer of knowledge from its point of origin to the point of opportunity[1] have crucial consequences for the working of the entrepreneurial economy.

The entrepreneurial economy gives a boost to economic development. Estimates by the Global Entrepreneurship Monitor show that something like 85% of the

T. Andersson et al., *Knowledge-Driven Entrepreneurship*, Innovation, Technology, and Knowledge Management 102, DOI 10.1007/978-1-4419-1188-9_9,

difference in economic growth between nations can be explained by the differences in environments conducive or detrimental to entrepreneurial activity.

The entrepreneurial economy replaces corporate capitalism with creative capitalism in which a new form of entrepreneurship becomes widely dispersed. Its main characteristic is the willingness to trade, communicate, and adopt innovation free from physical and cultural barriers, so a borderless context can lift an array of new ventures with a cosmopolitan and international focus from the start. The nationality of ownership will come to be viewed as almost irrelevant because of the strides made toward developing integrated and single markets (see Chap. 12).

Whether or not a society is entrepreneurial depends in part on the legitimacy and esteem accorded to those who pursue the entrepreneurial route. Communities in which entrepreneurship thrives create more jobs and wealth. Entrepreneurship foments the Schumpeterian process of creative destruction by which the new replaces the old. New opportunities are perceived, capitalized, and converted into marketable products or services.

Smilor et al. (2001), writing on the entrepreneurial process, offer this view:

"...entrepreneurs are the engines of our economy. They're the ones who are generating a lot of the innovations today and pushing the boundaries of technology – they are challenging the ways we act and behave and think. That kind of economic 'churn,' that kind of innovation, is so vital to our economy. If you look at leading economists who are studying the work on innovation, they see a direct link between the entrepreneur and the innovation process."

Red Herring's Jason Pontin has listed the following cultural attributes found in countries where entrepreneurship thrives:

"A literate middle-class population, a lack of admiration of aristocrats, a strong belief in risk-taking, little fear of failure, a love of the free flow of ideas (because dissent is not 'anarchy'), tolerance of diversity, belief in materialism, an 'infatuation' with technology, and a strong desire to educate one's children" (quoted from NCOE 2002).

Failure is an essential part of the entrepreneurial process. In this respect, there is a cultural divide between Europe and the USA. In the former region failure still stands as one of the biggest obstacles to achieving prosperity from innovation in an entrepreneurial economy. The anatomy of failure shows a losing process by which one persists in the failing ways, refuses to change the current course, and puts significant efforts into justifying that course. By contrast, in the American culture if one fails this means "getting blocked on," "backing out," and "restarting."

Exhibit 61: Entrepreneurship and the Entrepreneurial Economy

Entrepreneurship is the process by which knowledge inputs are changed into knowledge outputs – that is, into novel combinations of factors of production in order to offer new solutions to customers in the marketplace. This process requires passion,competencies, and a powerful skill to sell.[2]

An entrepreneur's passion for what he or she does is at the heart of the entrepreneurial process (Smilor et al. 2001).

It is the entrepreneurial spirit that nurtures the entrepreneurial economy. The entrepreneurial economy is a new age of capitalism, which serves the broader purpose of cultivating growth-oriented new ventures. It would make a difference to society in a reasonable period of time. The growing recognition by policy makers that what must be at the central ring of the country's performance for the years to come is the spontaneous optimism and motivation of nascent entrepreneurs to exploit their own specific talents and ideas in new areas of industry.

Leaders, Entrepreneurs, and Managers

There are three categories of individuals who take initiative and create the context on which the entrepreneurial economy thrives (see Fig. 12).

The first category includes the "catalysts," those who instigate entrepreneurship formation and renewal. They shape an environment of collaborative behavior and challenge the status quo and success factors. These individuals are acknowledged as the industry leaders. Today's "tyrants" and "titans" are those leaders who have been building a context of business collaboration and creating a spirit of common purposes and ambitions, while the new "heroes" are those who typically provide today's new visions whereby leaders undertake the consequent actions for the renovation (see the life cycle of leadership as it has been depicted in Fig. 13).

The second category features the "doers," those who build on the foundations established by the industry leaders. Their profile is that of the entrepreneurs who have not had the chance to gain a foothold on the industry-leaders' ladder. They are either owners of family companies or empowered managers and employees who are responsible for the company's entrepreneurial initiatives ("intrapreneurs" – as they have been termed).

Entrepreneurs constitute the glue or the "jam in a sandwich" between leaders and operational managers. Their major tasks involve attracting of new skills and resources for supporting trust-based interfirm relationships, endeavoring to achieve continuous performance improvement needed to nurture the spirit of common ends and aspirations, and the creation and pursuit of opportunities in accordance with the new ideas the leaders have envisioned.

"Developers," the third category, are those who transform the ideas of the leaders into concrete proposals, which the entrepreneurs then convert into reality. The

Managers	**Linking dispersed resources and skills**	**Reconciling short-term and long-term commitments and pressures**	**Encouraging and supporting extroverted initiatives**	Task
	• Team builder	Ability to fine-tuning	Ability to pass responsibility on to subordinates, to empower	Skills
	• Knowledge of community of practices, understanding of interpersonal dynamics	Understanding short-term priorities as the means and long-term goals as the ends	Knowledge of the individuals' quality, understanding how to influence them	Knowledge
	• Attitude to integration and flexibility	Acuteness	Be a coach, a supporter	Traits
Entrepreneurs	**Attracting new skills and resources**	**Continuous performance improvement**	**Creating and pursuing opportunities beyond the clusters horizon**	Task
	• Ability to motivate and drive people	Ability to take care of demanding targets	Ability to grasp the new potential	Skills
	• Knowledge of skills and resources in and out of the cluster	Knowledge of the business	Knowledge of the changing times	Knowledge
	• Attitude to attract and engage	Spirit of competitiveness	Creative, intuitive, eager to challenge the unknown	Traits
Leaders	**Building a context of strong trusting relationships**	**Creating in the cluster a spirit of common purposes and ambitions**	**Challenging the embedded success factors by a new vision**	Task
	• Inspiring confidence and creating beliefs	Ability to communicate	Questioning and demanding	Skills
	• Knowledge of the business culture, structures, processes	Broad knowledge of all players in the business	Understanding of how doing new things in new ways	Knowledge
	• Fairness	Insightful	Visionary-minded	Traits

Fig. 12 Leaders, entrepreneurs, and managers: their tasks and attributes

developers' profile is similar to the operational managers, who link resources and skills attracted by the entrepreneurs, develop the initiative for the opportunities that the entrepreneurs have opened up, and align short- and long-term commitments on the basis of the entrepreneurs' performance metrics.

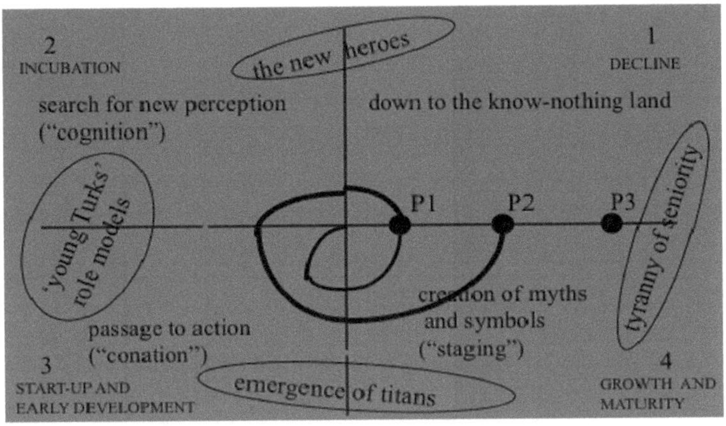

Fig. 13 Life cycle of leadership.
The life cycle of leadership can be traced through four situational patterns – namely: down to know-nothing land, search for new perception ("cognition"), passage to action ("conation"), and creation of new myths and symbols ("staging"). There is an enduring conflict between the old guard management with its tyranny of seniority and the informality and openness of would-be heroes and drivers of change who steer the chaos. Anytime a community is confronted with the huge leap from tradition to innovation, the leadership of the past pioneers and early followers (today's tyrants and yesterday's creators of myths and symbols) still remains and influences the community. New heroes suffer from a limited understanding of their presence. "Young Turks" are in search for a new perception, tapping creativeness and benefiting from diversification inside the traditional industries or moving on a new pasture – for instance, by attempting to carve out a niche for themselves in the Internet economy. They set up new ventures in close collaboration with tech-savvy, experience-poor new entrepreneurs. Among their backers there are financiers, venture capitalists, individual consultants, and consultancy firms

Notes

[1] According to the "Second Law of Knowledge Dynamics" that states that "value is created when knowledge moves from its point of origin to the point of need or opportunity" (see Chap. 2).

[2] Interviewed by Joelle Wolstein, Ray Smilor, President of the Foundation for Enterprise Development, La Jolla, CA, argues that selling is a powerful skill that every entrepreneur has to have. He comments: "Really great entrepreneurs first sell themselves to investors and others, then they sell the concept of their company to employees, then they sell their product or service to customers. And really [great] entrepreneurs begin to see selling not as this hated skill, but as a skill that's essential to their success and the success of their company. A way to make selling less painful is to see it for what it really is. You're not forcing something on a person that they don't want; you're helping somebody buy what they really need. So the first step is to change your frame of mind about the skill of selling. Second, I think it's important [to be an] exceptional listener — to take time to listen to what the other party has to say, to what their genuine need is, to what the problem is. Then shape a response that solves the problem or meets the need. Third, I think [you need to] actually practice selling – get out there and get better at listening, responding and probing – those kinds of things make this most hated skill become quite natural." Source: Ray Smilor, "Why you do what you do as an entrepreneur — and why we need you," *Entrepreneurs.com*, June 7, 2001.

Chapter 10
Types of Entrepreneurs

For the duration of its collective life, or the time during which its identity may be assumed, each class resembles a hotel or an omnibus, always full, but always of different people.
(Joseph A. Schumpeter)

An entrepreneur is an independent agent who adopts a set of rules, consistent with a "search-and-satisfying" type of behavior, in order to reach goals such as the growth and profitability of his or her company. In doing this, curiosity and an instinct for exploration drive the entrepreneur – a combination in which intentional action and the faculty of making lucky and unexpected finds by accident sit side by side.

Somehow, entrepreneurs are the incarnation in the real world of the characters of a fairy story, "The Three Princes of Serendip,"[1] who "were always making discoveries, by accidents and sagacity, of things they were not in quest of." They accept that everything is a matter of degree – that is, they "expect every 'well-formed' statement to be not true or false, but true more or less or false somewhat." In other words, they believe that "A and not-A holds to any degree" (Kosko 1994). By the way, this logic, which is of Buddhist origin and has been defined as "multivalent" or "fuzzy," in contrast to Aristotle's "bivalent" logic, marks the most distant frontier between entrepreneurs and those scientists, mostly from the Western culture, who deem that fuzzy logic is wrong and pernicious, notwithstanding the number of innovations that originated from it.

Therefore, to find their own way, entrepreneurs look for many windows and take quick decisions, albeit sometimes they need to move slowly toward their target. Eventually, they have to see which way the wind is blowing.

Many myths surround entrepreneurs, such as those concerning the high level of risk that presents a new undertaking, the importance of the business plan, and the reliance on venture capital. Whereas it is said that entrepreneurs take wild, uncalculated risks to start their companies, in reality many entrepreneurs when they start out do not have much to lose, and they have an uncanny ability to convince others – employees, individual investors, suppliers, and landlords – to share their start-up risks. It is also a common view that entrepreneurs have a well-researched, well-conceived strategic plan when they start their companies. Vice versa, what enables most entrepreneurs to be successful in their new ventures is their flexibility to

T. Andersson et al., *Knowledge-Driven Entrepreneurship*, Innovation, Technology, and Knowledge Management 102, DOI 10.1007/978-1-4419-1188-9_10,

change. Starting a new business is like jumping from rock to rock up a stream rather than building a bridge from a blueprint. Companies develop tightly constructed business plans when they are ready to seek outside investments. That all entrepreneurs rely on venture capital to fund their businesses is a third myth. Venture capital companies only fund a very small number of businesses each year – about 4,000 overall in the USA. Most entrepreneurs start with their own money and money from friends and family and only look to venture capital when they need to capitalize on their successes, usually in the later stages of growth (US National Commission on Entrepreneurship 2001b).

Entrepreneurial Opportunity and Capacity

Entrepreneurship flourishes at the crossroads where entrepreneurial capacity meets *entrepreneurial opportunities*. To take advantage of an opportunity (i.e., to transform a possibility into a business reality), individual commitment and capability are primary conditions. Commitment implies motivation, and capability calls for entrepreneurial attributes. Therefore, the *entrepreneurial capacity* embraces motivations and attributes.

Should the availability of entrepreneurial capacity exceed opportunities, unfulfilled expectations would discourage an otherwise positive attitude toward entrepreneurship. Similarly, entrepreneurial opportunities in excess of capacity will not generate commercial outcomes. Both imbalances depress the level of entrepreneurial activity (see Fig. 14). Consequently, the growth of entrepreneurship is the product of a three-pronged strategy:

- Developing entrepreneurial capacity
- Cultivating and detecting entrepreneurial opportunities
- Raising the level of equilibrium between entrepreneurial capacity and opportunities from the creation of small firms to that of entrepreneurial growth companies

Exhibit 62: Entrepreneurial Opportunity and Capacity

"An entrepreneurial opportunity consists of a set of ideas, beliefs and actions that enable the creation of future goods and services in the absence of current markets for them" (Acs and Varga 2004)

Entrepreneurial capacity consists of motivation and entrepreneurial attributes.

Motivations

- Capacity to think for oneself
- Self-confidence
- Sense of autonomy, independence, and risk-taking
- Intense emotions

(continued)

Exhibit 62 (continued)

Entrepreneurial Attributes

- Clarity of leadership
- Openness and inquisitiveness that stimulates innovation and learning
- Creation of new value or organizational capability
- Flexibility to change
- Relationship-building skills
- Ability to convince others (employees, individual investors, suppliers, and landlords) to share start-up risks

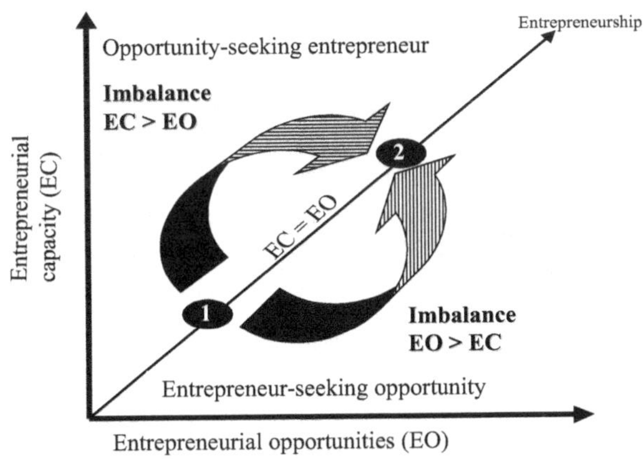

Fig. 14 Matching entrepreneurial capacity and entrepreneurial opportunities

Entrepreneurship in the Knowledge Domain

An abundance of talented people does not automatically produce a high-performing community in entrepreneurship. World-class researchers are not often world-class knowledge-driven entrepreneurs.

To encourage entrepreneurship in the knowledge domain, it makes sense investing both in raising entrepreneurial education among individuals with a scientific background as well as among people with a managerial and humanistic background and melding these groups together. From the self-employed to people working within organizations there is a diverse range of individuals displaying entrepreneurial attributes. Disparate factors feed the process of entrepreneurship in the knowledge domain, among which are the following:

- Cultural roots, heritage, knowledge, skills, and experience enjoyed by individuals. They hold a greater value than capital equipment.

- Entrepreneurial spirit among youth, enhanced by an entrepreneurial education and the process of learning about running a business at a young age.
- Students and graduates from business and technology schools, from science and art faculties, who are prepared to cooperate in the setting up of new ventures.
- Business agents who allow a better matching between scientists and technologists with product and process ideas, and managers, financial and marketing specialists able to commercialize those ideas. Moreover, business agents are focused on marketing and sales, and, therefore, they maintain daily contacts with growing businesses.
- Mental tools for exercises in market foresight. For instance, time-to-market techniques and market-focused technology audit of research activities in order to seize the window of opportunity. As experts say, "Coming too early the spin-off burns too much cash. Coming too late others have already taken up market positions."
- Partnerships between would-be entrepreneurs and established companies as "launching customers." In particular, thanks to links with companies operating in different industries; start-ups can take a broader view of what they are going to have to learn – which frequently come from outside its own industry.
- Networks made up of business firms experienced in market foresight, local seed capital funds, and wealthy individuals ("business angels"). This type of collaboration increases both the amounts of funds to develop a sufficiently large portfolio of pre-start-up organizations and the ability to transform inventions into marketable innovations. The thicker the networks, the richer the opportunities the new businesses can exploit.

Exhibit 63: Market Foresight

Market foresight is the ability of discovering future markets by developing sustainable future visions and not merely projecting forward forecasts, whereas market watch is the ability of matching innovative solutions with problems detected in the actual markets.

Exhibit 64: Business Angels

Important innovations are lost due to the lack of management support, a proper market approach, and risk-bearing capital.

Experience has proven that quite a number of projects remain forever buried in research; commercially promising technologies often get stuck in an early stage, and many of the the firms that are growth-oriented fail in their early lives because they cannot cross the so called "valley of death" or "no man's land," due to the lack of a proper market approach and risk-bearing capital.

Business angels are individuals, oftentimes former entrepreneurs, who invest first and foremost in entrepreneurial people for a variety of reasons, ranging from

(continued)

Exhibit 64 (continued)

fun to making a high return on investments to an interest in mentoring promising, fast-growing start-ups. In addition to providing funds, they provide advice, counsel, and contacts to new entrepreneurs.

Typical angel investments fall in the range of $50,000 to $500,000 per transaction. This relatively small amount of money fills an important niche in private equity markets. Entrepreneurs regularly report that obtaining funding in the range between $100,000 and $2–3 million can be quite difficult. These amounts are often too small to interest venture capital firms or banks, but are too large to be generated via friends, family, and other personal contacts. Thus, angels play a critical role in nurturing entrepreneurship ((see the Amazon's case in Fig. 15).

Besides money, angels' expertise yields benefits to start-ups in terms of the following:

- Supporting potential entrepreneurs to draw up business plans and business strategies in the field of market analysis, feasibility studies, financial planning, commercial contacts, recruitment of additional management.
- Providing advice and support in bringing new products or product prototypes into a commissioning phase.
- Involving professionals and businessmen abroad so they can act as spearheads for the newborn companies to enter foreign markets.
- Bringing the business idea to a stage at which it will be an interesting subject for investment by a venture capital organisation.organization.
- Making available an incubator-type organisation organization equipped to host growth-oriented start-ups.

All in all, business angels help transforming a workforce- and payroll-based economy into an entrepreneurial one that enhances business partnership contracts. They also contribute to make the economy as a whole more productive and growth-biased, since the wealth effect generated by a broad participation in fast-growing start-ups encourages consumption.

Traditional business angels worked alone or in small, unorganized groups. Today, organized angel groups have become a key component in the private capital markets. These formal groups may actually serve to encourage more angel investment. By reducing exposure and formalizing the investment review process, formal angel groups may serve to encourage more people to consider investing.

As organized groups have emerged, they have also taken on more diverse orientations. In the USA, angel groups targeting women or ethnic minorities are also growing. Angel networks also appear to be sprouting in rural areas and other regions outside of the traditional technology hot spots.

Formal procedures and industry "best practices" have accompanied the rise of the business angel groups. Most of the groups seem to use similar investment screening procedures. At the same time, groups still utilize a wide range of organizational and legal structures.

Source: Excerpts from US National Commission on Entrepreneurship 2002b

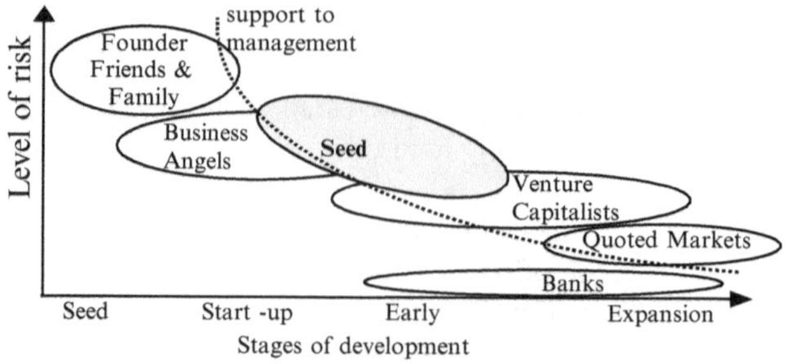

Stages
Seed – incubation; concept to be proven and developed.
Start-up – product developed; initial marketing takes place.
Early growth – first enlargement; the company is yet unprofitable.
Expansion – fast enlargement; the company is profitable, goes public (that is, stock exchange quotation) or is sold by the financiers.

Business angels provide a strong support to management, which drops heavily in the later stages (see the dotted line). Seed funds bridge the gap between informal investors and the venture capital industry.

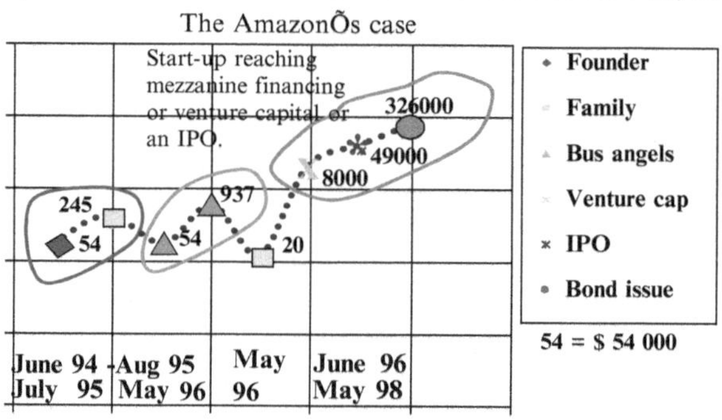

Fig. 15 Financiers of growth-oriented start-ups

Creative Entrepreneurship

Creative entrepreneurship discerns new business paradigms through which opportunities are pursued that break away from the familiar routines. Creativity is stimulated by investment decisions whereby the received wisdom of corporate principles of rational calculation is counterbalanced by the capability of the "animal spirits" to seize mere ideas. "If the animal spirits – Keynes noticed in his General Theory – are dimmed and the spontaneous optimism falters… enterprise will fade and die," and nascent entrepreneurs will be disheartened in the motivation to exploit their own specific talents and ideas in new areas of industry.

Creative entrepreneurial thinking needs to be supported by a strong culture of commercialization. Once a new thought has been developed, it must be validated, and then a prototype offering has to be created, the competitive environment assessed, the offering tested, and feedback used to refine the offering. This business process leads to a business plan tightly constructed and executed when the new entrepreneur is ready to seek outside investors.

Creative entrepreneurship makes difference in the entrepreneurial vibrancy under three perspectives: creativity in technology, creativity in product planning, and creativity in marketing.

Creativity occurs when both technology impact and market impact of a given undertaking is high. Creativity inflicts major changes on the core competence or

Exhibit 65: Akio Morita's Three Creativities

S [Science] does not equal T [Technology] and T does not equal I [Innovation]: *this* is the title of a famous lecture by Akio Morita, Sony's founder, at the Royal Society in London, in 1992.

The progress in business knowledge is relevant as much as developments of science and technology. From Morita's perspective, "just having innovative technology is not enough to claim true innovation." True innovation is made up of three key elements, which Morita calls the "three creativities":

- Creativity in technology
- Creativity in product planning
- Creativity in marketing

Contrary to the common belief, creativity in technology, or technological clairvoyance, is far from enabling technology entrepreneurs to succeed. Technology, even a good one, does not sell itself.

Creativity in product planning, argued Morita in that lecture, "is so important, though many do not seem to recognise this… What difference does it make how fantastic and innovative your technology is if you do not have the ability to design a useful, attractive, user-friendly product?"

"Videotape recording technology – Morita observed – was first introduced to the consumer market in 1965, but the home video market was not born until 1975. That was when innovative product planners took the tape out of the reels and put it into a convenient Betamax cassette for home use.

Creativity in marketing also cannot be overlooked. Again, if you have great technology and even a great product, you will only find success if the market is informed enough to welcome your product."

Borrowing an example from Sony's history, Morita made reference to the case of the Walkman. He submitted that many have called it an innovative marvel, but where is the technology? All components to make it were already available on the shelves. "Frankly, it did not contain any breakthrough technology. Its success was built on product planning and marketing" (Morita 1992).

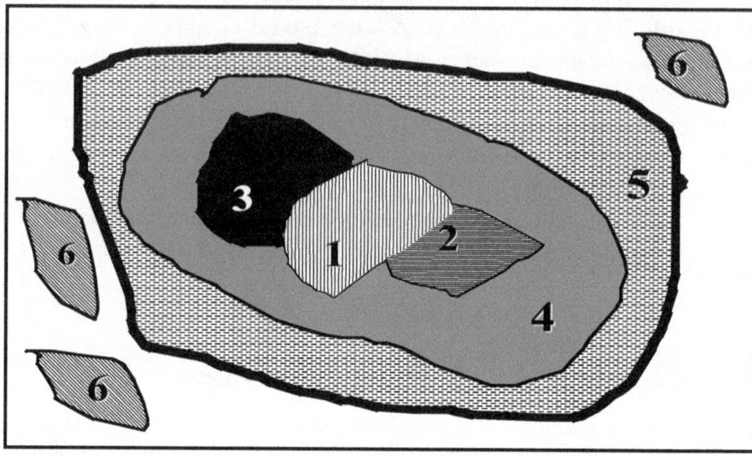

Fig. 16 Creativity map: regions of business knowledge

even induces the abandonment of what the company until then thought it was good at.

Four different situations emerge from the intensity of both the technology impact and the market impact caused by a given innovation on the company to which it is coupled. In turn, the company is articulated in six regions of business knowledge, as shown in Fig. 16.

- The core competence or domain expertise (i.e., what the company is good at – region 1)
- Known unknowns (region 2)
- Unknown unknowns (region 3) nearby the core competence
- Knowledge of the current competitors (region 4)
- Knowledge of outsiders as potential invaders (region 5)
- Perception of original technologies from today's new scientific discoveries that can turn into tomorrow's markets: "technology islands" (region 6)

Unlike creative entrepreneurship, scientific prone clairvoyance is a distinctive trait of pure scientists and researchers who look ahead, beyond the frontier of the known domain. A prevailing sentiment of a manifest destiny nurtured by a rugged individualism urges them to traverse the Pillar of Hercules beyond the frontier of the today's world of knowledge. Their actions help establish pioneer settlements at the extreme borders of the business world. These are "technology islands" whose time is not yet ripe for their commercial exploitation.

Exhibit 66: Creative Entrepreneurship vs. Science-Prone Clairvoyance

Clairvoyance exhales a flavor of science. Creativity secretes a business touch of geniality.

Pure scientists and researchers yearn to go beyond the utmost limits of the current knowledge domain. They have a long-term commitment to solve problems that appear impossible.

Exhibit 67: Time-to-Market and Routine

Time-to-market and routine are the threads that connect efforts companies make to get success from advancements within the core competencies to the resolution with which they carry on the building of a new business order.

Time-to-market encompasses the notion of going the pace in the introduction of product or service innovation improvements to the marketplace.

Routine is the ability to work with method to gain in efficiency. Compared to routine, time-to-market can have a larger effect on profit than on product cost.

Exhibit 68: Improvement

Improvement is not innovation. Improvement is about the amelioration of the status quo; innovation is about the disruption of the current state of the art, doing something in an entirely different way (Horibe 2002).

By helping current businesses release resources to be employed in creative ventures, time-to-market and routine operations are leverages for creative entrepreneurship.

The action of monitoring gaps that are known (such as, articulated user needs) and discovering unknown drawbacks (such as, the behavioral needs of potential customers) makes time-to-market suitable to contribute to go more quickly and more effectively to markets with product or service innovation improvements resulting from ready-to-use technologies.

Routine, on the other hand, enables constant advancements through continuous improvements upon existing models.

Creative Entrepreneurs: Technology Entrepreneurs, Knowledge Entrepreneurs, Entrepreneurial Scholars, and High-Expectation Entrepreneurs

Creative entrepreneurs uncover new business developments that need fresh resources either in money (such as, risk capital) or in kind (such as, human capital – that is, knowledge and competencies to provide solutions for customers).

Creative entrepreneurs are essential to the mastering of a new domain. They pave the way for the economic success of their community. In the business-as-usual, good character actors replace the protagonists, but the shape of things to come is traced by the emergence of new leading personalities who are willing to change the fabric of traditional mental habits and conventional ideas supported by people with similar thought processes. In Appendix 10 we highlight the cases of Enzo Ferrari and Lorenzo Del Vecchio who have been so instrumental to the renewal of the post-war industrial fabric in Italy.

Exhibit 69: Features of Creative Entrepreneurs

- Hedgehog-minded personality or versatile talents (fox-minded individuals)
- Restless
- Ability to organize and maintain a clear view
- Ability to get into a market as quickly as possible
- Ability to connect and integrate
- Awareness of their worth, having started from scratch and knowing that results are obtained with struggle and diligence, not by chance or shortcuts
- Awareness of the march of time

Technology Entrepreneurs or Technopreneurs

Technology entrepreneurs or "technopreneurs," knowledge entrepreneurs, and entrepreneurial scholars turned into intellectual venture capitalists typify the knowledge-based, creative entrepreneurship. They are, usually but not always, scientists and researchers, who, being accustomed to work with industry, combine research, creativity, entrepreneurship, and willingness to take calculated risks.

However dazzling the technology may prove to be, addressing the right window of opportunity into the marketplace will be the dynamo powering creative entrepreneurs to international preeminence. By implication, they must be endowed with perception of the significance and nature of events before they have occurred, care in providing for the future and the ability to look forward: in one word, they need market foresight. Technopreneurs who are lacking in such foresight are "Men who learn only through suffering" – as Demosthenes warned in his work *On the Trierarchic Crown.*

A *technopreneur* is a person who effectively brings together research talent, venture capital, new business concepts, and management skill in order to create commercially successful technological innovations or, alternatively, uses technology to effectively leverage innovations (Milton-Smith, 2002). The technopreneur's *modus operandi* consists precisely in combining the Morita's "three creativities" (See Exhibit 65).

The conditions for survival and growth of technology-based entrepreneurship are heavily influenced by the technopreneur's attainment of managerial competencies, his or her networking capability in the business environment and, in particular, the development of both market relationships and embedded ties (i.e., close or special relationships) with large companies.

Exhibit 70: Technopreneur

The word "technopreneur" is of recent origin. It was probably first used in Singapore in the early to mid-1990s as a way of highlighting the need to generate more entrepreneurship in the technology sector, as opposed to the more traditional areas in which entrepreneurs operate, such as property development, retailing, and trade. There was certainly a widespread view that technological entrepreneurship in Asia — not just Singapore — had lagged behind that in Europe and the USA. Thus, in recognizing the need to spur technopreneurs, the Malaysian government launched the Technopreneur Development Flagship programme in November 2001, and the Technopreneurs Association of Malaysia was established.

According to the Technopreneurs Association of Malaysia, "Technopreneur" means a person who uses the application of digital technology [that is, an economy based on the digitization of information and the respective information and communications infrastructure including the technology therein involved] to industrial or commercial objectives or one who organizes, operates, and assumes the risk for a business venture in the Digital Economy.

Source: Milton-Smith 2002; http://www.technopreneurs.net.my

Technopreneurs:

- Have familiarity with the academic world
- Seize opportunities for launching start-ups in technologically progressive new firms enjoying sustained fast growth, mostly in the fields of ICT, electronics, computers and software, biotechnology, life sciences, and environmental technologies
- Have a technical-biased culture pushing their ventures into a tiny niche market where the survival rate is low
- Lack in a business vision and do not possess the adequate knowledge of competitive forces in the market

Moore (1999), Chairman of the Chasm Group and a venture partner at Mohr, Davidow Ventures, has drawn up a model that has much to say about forward signals that help technopreneurs to bridge the gap or "chasm" between early adopters of a new-brand technology and pragmatic buyers in order to reach the lucrative mainstream market (Fig. 17).

Technology-based start-ups strive to create high-tech products tied to the spirit and culture of their potential buyers. So, the solutions they offer have to be not only functional and convenient, but also stylish and attractive. Technopreneurs' products must be functional material objects and they must appeal to the human spirit. In this sense, one can say that techopreneurs operate in a culture industry (Moritani 1982).

In doing so, they need early adopters gaining experience from those solutions. Since most novel technologies –Moore argues – can offer only partial solutions, a new market, he further states, can be disclosed only by enthusiasts or visionary customers – that is, by early adopters ready to accept a very high pain in exchange for a very high gain provoked by an innovative but imperfect response to their requirements. Early adopters overcome "self-incompatibility," as Liebowitz (2002) has defined that kind of lock-in dependence, which arises from the products or services already in use. Self-incompatibility arises in consequence of both the costs of learning how to use a new product and the difficulty of using it alongside other products the adopters already own. On the other hand, they are not affected by the "external incompatibility" – that is, by the fact that buying the innovative product they make a choice incompatible with the preferences of the majority of consumers. The latter prefer to continue with the most familiar product because the rival does not look much better in terms of advantages and disanvantages that its use entails.

In contrast to those visionary customers, who are keen to make a quantum leap forward, pragmatic buyers constitute the early majority looking for such an improvement in the new product that makes it a complete solution through which they can get a very high gain while suffering only a modest distress. The vendor, who fails to get the remaining *x*% in order to deliver a whole product, will be falling

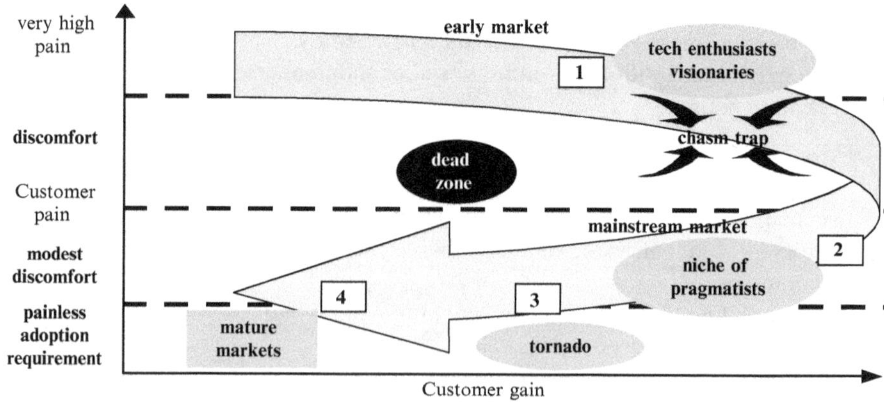

Fig. 17 Moore's model of terchnology adoption life cycle
Source: Adapted from Moore 1999

into an abyss. Technopreneurs, therefore, have to be very attentive in decoding the signals transmitted by the pragmatic buyers "who look to each other for guidance during the chasm phase": if "no one like them is adopting, so they too hold back" – Moore maintains. Correctly interpreting those signals means that technopreneurs have been focusing on a niche of pragmatists as a chasm-crossing mechanism.

Exhibit 71: Visionary Customers and Pragmatists

Visionary customers and pragmatists are comparable to lighthouse keepers who, by means of gain (green light) and discomfort (red light) innovative products and services cause them, transmit signals to technopreneurs navigating in technology markets.

Once in the mainstream market, those technology-based start-ups committed to make available a painless adoption with a high gain can capture the benefits of the market takeoff: what Moore calls the "tornado of an hypergrowth phase at triple-digit rates for several years." Finally, the innovative product or service being utterly successful (a "must-have" item), the tornado's vortex will change into the calm of a mass-adoption market: a mature phase that rewards customers with a pair of very modest gains and no discomfort.

Many vessels are expected to be in trouble navigating in such dangerous waters. Not only the bulk of new ventures fall into a chasm trap, but even those crossing the chasm can hit the sandbanks. What they offer might appear attractive to would-be customers but not still a must. Overall, a small gain will accrue to buyers in exchange for modest pain. In a "dead zone" companies cannot survive. Either they endeavor to change their current course or go out of business.

Today's technopreneurs resemble a modern day Candide in search of the best-of-all possible worlds. They place emphasis on the upsurge of markets that compensate buyers with very high gain and no distress. How many technology-based start-ups will be keeping or breaking their promises depends on the mode of navigation in the dangerous waters vividly traced in Moore's model.

Knowledge Entrepreneurs

The move from industrial agglomeration to knowledge agglomeration brings along the reformulation of domains and criteria of business environments, operating rules, and leadership. New life cycles of contents shape innovative formats and spaces, where knowledge flows are exchanged in between networks of ubiquitous players. Hyperspeed connections accelerate questions and answers, thus boosting interactions across variety of fields. XXI century dynamics lead to emerging knowledge-market forces and open up to new positions for new business leaders: the knowledge entrepreneurs.

Knowledge entrepreneurs (KEs) are knowledge creators and knowledge processors. They are "those who create value in the borderless knowledge markets through the infinite resource of knowledge that they put into action with the purpose of making advancements in the society, the economy, and the environment" (see Chap. 1).

Founders of companies intensely based on knowledge are embedded in borderless knowledge markets. The cross-border strategy pursued by KEs makes them agents of cultural integration.

The fact that one of the scarcest resources in a knowledge-based economy is the organization's ability to create new knowledge makes KEs adept at tapping into the global talent pool of the creative class that includes professionals, artists, musicians, scientists, economists, architects, engineers, and managers whose features are talent, high skills and education, creativity, capacity of innovation, and entrepreneurial spirit. So, KEs contribute to raising both the index of cultural integration and the level of creative employment.[2]

Exhibit 72: Knowledge Entrepreneurs

- Trust builders
- See creativity of staff and the capacity of innovation as the primary production factors
- Treat customers as partners and knowledge workers as revenue creators
- Manage the environment in which knowledge is created
- Create informal networks of alliances: Knowledge entrepreneurs rarely act alone. Entrepreneurs are traditionally seen as individualists. Knowledge entrepreneurs usually emerge from a network of complementary ideas and people (Leadbeater 2000)
- Forge and handle relationships that are sideways (i.e., where there is no authority and no orders)
- Tear down the man-made barriers (cultural, institutional, and geographical) that prevent knowledge sharing
- Build bridges to different communities and countries (Sveiby 2000).

Green entrepreneurs ("ecopreneurs") introduce eco-friendly (or relatively more eco-friendly) products and processes into the marketplace. Knowledge-driven business entrepreneurs create viable and growing business organizations in the knowledge markets. Social entrepreneurs accomplish social purposes, in addition to being commercially viable. They focus on innovations that have social impacts. What the aforementioned forms of entrepreneurship have in common is that their activities shift toward knowledge-intensive processes.

(continued)

Exhibit 72 (continued)

The Knowledge Entrepreneurs Network YEAM: Young European Avant-garde Minds

While in search of knowledge trends, knowledge entrepreneurs anticipate knowledge agglomeration developing and leading their networks, as the YEAM case.

YEAM – Young European Avant-garde Minds – is an open network born in 2006 in the hearth of Europe. YEAM aims to design a distinctive hub for the emerging creative entrepreneurs, attracting transdisciplinary talents and clustering them around a knowledge pool in which intents enhance actions. Here, knowledge is embedded in individuals; and individuals are the key elements of the knowledge pool. This creative and innovative knowledge mobilization network leads to increasing the impact of the research full scale, exploiting unique R&D themes and potentialities and inflowing entrepreneurial resources by attraction mechanism.

Large, diverse, and multidisciplinary concentration of knowledge and talents can develop from those leaders further, exploiting alternative paths of knowledge mobilization, starting from their intellectual capital.

Entrepreneurial Scholars

Entrepreneurial scholars have a relatively clear sense of the probability of a successful commercial outcome from their curiosity-driven research – and their research evolves into business-driven, goal-oriented work. This evolution results in both a paradigmatic shift achieved by the adoption of a new intellectual model and a phase change necessitated by the transition from research to entrepreneurship (Carayannis and Formica 2006).

Knowledge entrepreneurs and entrepreneurial scholars who have turned into intellectual capitalists open up new perspectives for outsourcing innovation.

If the supply of KEs is low, the outsourcing of innovation is a decision that must be taken within a constrained vision – simply that of a tangible-assets-intensive process controlled by companies making outsourcing decisions. Those companies focus on what *they know they do not know*. Under these circumstances, outsourcing decisions keep to chartered waters: navigation depends on knowing how to keep innovation-induced pressure on tangible assets under control.

In contrast, an abundant supply of KEs encourages intangible-assets-intensive processes, whereby companies making decisions for outsourcing innovation "learn" rather than "control." In this case the focus is on what companies do not know they do not know.

Exhibit 73: Entrepreneurial Scholars

They are, usually but not always, scientists and researchers, who, being accustomed to work with industry, combine research, creativity, entrepreneurship, and willingness to take calculated risks.

High-Expectation Entrepreneurs

Once technology advancements met entrepreneurship then market novelties are created over time and turned into high-expectation start-ups, which are firms launched by high-expectation entrepreneurs with the aim of significantly growing their companies. These firms pursue the commercialization of an innovative new process, product, or service. That encounter is a force that tends to push the economy away from a state of equilibrium. In fact, high-expectation entrepreneurship provokes disequilibrium, which is a state of change-induced imbalance with no tendency to stasis. We can say that high-expectation entrepreneurship is high-impact entrepreneurship – the kind that drives the growth of technology industries and one of the factors that contribute to shape the economy as an open, complex adaptive system.

Exhibit 74: Technology Meets Ambition: Ryanair

An outstanding example of the growth that can occur when technology advancement meets entrepreneurship is that of Ryanair. At one point a small struggling Irish regional airline, its oversized ambition began to be realized when a business innovation it copied from Southwest airlines intersected with the emergence of the Internet and Internet reservations systems, which dramatically reduced the transaction and distribution costs associated with passenger reservations. In this case Ryanair was able to adopt Southwest airline's tried and trusted model of point-to-point flying, coupled with a single aircraft type fleet to minimize fleet total cost of ownership while flying into smaller regional airports with lower landing charges.

While this model was successful Ryanair was limited by the ability to scale its own call center to handle expansion of its business. The arrival of Internet and Internet reservation systems enabled Ryanair to scale new routes and quickly improve yields through a ubiquitous Internet interface and computer-based yield management systems. As Ryanair introduced new routes, passengers had immediate and easy access to low fares and a network-type effect was created. Almost overnight Ryanair was seen as a transformative force in European airspace, creating a medium for not just low-cost holidays but more importantly also a platform for enabling entrepreneurship in Europe as travel costs associated with business meetings became much more affordable enabling new physical connections and meetings to take place, creating new options, and enabling new business.

(continued)

Exhibit 74 (continued)

These kinds of physical interactions could not have taken place in the prior prohibitively high airfare environment that existed. The fact that at one point Ryanair's market capitalization surpassed the total value of Lufthansa and British Airways combined shows the power of high-expectation entrepreneurship. It is important to note that prior to the success of Ryanair, multiple different business model experiments were tried, which burned real capital.

The presence of an experimental lab environment, which could simulate different business models taking into account consumer behavior, regulatory changes, and other environmental factors, perhaps could help similar high-expectation entrepreneurial activity avoid some expenditure and accelerate the path to profitability.

High-expectation entrepreneurship deserves special focus because of its oversized impact on economic growth. According to the Global Entrepreneurship Monitor, less than 7% of nascent entrepreneurs expect to employ 50 or more employees within 5 years; however, the economic impact is disproportionally positive as high-expectation entrepreneurs are responsible for up to 80% of total expected jobs by all entrepreneurs. Also as an example for a group of highly developed countries a 1% increase in the general rate of entrepreneurial activity raises economic growth by 0.11% while a 1% increase in high-growth entrepreneurship yielded two-times multiplier effect with a 0.29% increase in GDP growth (Stam et al. 2007).

High-expectation entrepreneurs exhibit oversized ambitions that mold unpredictable growth patterns: from exponential and oscillating to declining and collapsing trends. As long as those entrepreneurs inject in the economy free energy, which corresponds to the flow of information and human interactions in view of new venture creation charged with high expectations, the system remains open to all possible states: sometimes being near to equilibrium or in an equilibrium-like state, some other time exhibiting a position too far from equilibrium. Additionally, a variety of management responses to disequilibrium states can lead to underdamped, overdamped, or critically damped management responses due to the associated uncertainty and likely high velocity of the venture. Moreover, complexity and adaptation emerge as significant characteristics of the economic system insofar as all agents involved in the creative process of high-expectation new ventures interact and adapt to each other and the context in which they are embedded.

Purposeful, high-expectation entrepreneurs explore with new eyes the uncharted territories of unforeseen circumstances and undiscovered opportunities. From the scratch they grow new markets, which are evolutionary organisms effective at innovation. Complex problems they tract get empirical validation in laboratory experiments where the function and performance of high-expectation start-ups are evaluated. The results of experiments give entrepreneurs, financiers, and policy makers a deeper understanding of the actual workings of real-world new markets.

Experiments point out how high-expectation entrepreneurs should cultivate market outcomes, which behavior should guide trust building between the former and their potential financiers, and how policy makers should design and test "rules of the game." Persistent beta states for the business model and underpinning venture offerings become the norm. Rapid experiment iteration and rapid solution prototyping go hand-in-hand for the high-expectation entrepreneur, with plateaus of stability introduced to the iteration cycles, to enable commercialization and value capture from the evolving offerings.

Exhibit 75: High-Expectation Entrepreneurs

High-expectation entrepreneurs account for just 7% of global start-up activity. However, they make a disproportionately large contribution to economic prosperity and job creation. The high-expectation entrepreneur is typically a young male, has a higher education, comes from an upper-income household, and has little fear of failure.

According to a report from an international audit and advisory organization:

1. Education and household income, as well as entrepreneurial activities and attitudes, were significantly associated with high-expectation and high-growth entrepreneurship. High-expectation and high-growth entrepreneurs had a higher level of education than other entrepreneurs and the general population.
2. Only 30% of all categories of entrepreneurs were women, whereas less than 25% in the category of the high-expectation and entrepreneurs were female.
3. An individual's decision to launch a new venture is affected by both the environment, and his or her personal characteristics and skills. "It's a combination of these two elements that determines whether a particular opportunity has potential for growth in the eyes of any potential entrepreneur," said Hilton Saven, senior partner of Mazars Moores Rowland, the South African arm of the Mazars group.
4. There is a sharp division between early-stage, high-expectation entrepreneurs and their already established high-growth counterparts. Almost 20% of the early-stage group were between 18 and 24 years old, while only 3% of the established group fell into this age bracket, with more than 50% over 45 years old.

Source: Report released by accounting and advisory group Mazars, in collaboration with the Global Entrepreneurship Monitor. The report was based on interviews with 678,714 adults spanning 53 countries over a period of 6 years, making it the largest study of high-growth entrepreneurship yet conducted. It provides important clues as to the make-up of the high-expectation entrepreneur. Quoted from Sanchia Temkin, "Entrepreneurs not solution to unemployment," Business Day – News Worth Networking, Saturday, August 16

Preparing the Innovating Entrepreneurs of the Future

Education levels are a powerful predictor of entrepreneurial achievements and interest in high-growth entrepreneurship. In the USA, companies founded by entrepreneurs with graduate degrees grow fast and employ at least 20 people within 5 years. Founding members of R&D-intensive firms tend to be linked to and source knowledge from universities. Hence, investments in entrepreneurial human capital are vital to the successful creation of new firms. Strong public funding of education and university-level research, a vigorous technology-transfer mechanism, and respect for intellectual property rights strongly contribute to spur the entrepreneurial spirit and enhance entrepreneurial performance.

Exhibit 76: Entrepreneurship Education: The American Way

Entrepreneurship is a developmental process. We recognize the importance of nurturing the entrepreneurial spirit from early ages and continuing it right through all educational levels. In most cases entrepreneurship is infused in classes where it provides the context for learning other basic skills and motivating students to want to learn. In the more advanced grades it also has become a separate course supporting the outcomes of the higher levels of the lifelong learning model.

Entrepreneurship education means many different things to educators – from primary schools to university, from vocational education to a university MBA. At each level of education, it is reasonable to expect different outcomes as students mature and build on previous knowledge. But the overall purpose remains to develop expertise as an entrepreneur.

Entrepreneurship is a lifelong learning process that has at least five distinct stages of development. This lifelong learning model assumes that everyone in our education system should have opportunities to learn at the beginning stages, while the later stages are targeted at those who may specifically choose to become entrepreneurs. Each of the following five stages may be taught with activities that are infused in other classes or as separate courses.

Stage 1 – *Basics.* In primary grades, junior high and high school, students should experience various facets of business ownership. Motivation to learn and a sense of individual opportunity are the special outcomes at this stage of the lifelong learning model.

Stage 2 – *Competency Awareness.* The students will learn to speak the language of business and see the problems from the small business owner's point of view.

Stage 3 – *Creative Applications.* At this stage, students can take time to explore business ideas and a variety of ways to plan the business.

(continued)

Exhibit 76 (continued)

Stage 4 – *Start-up*. After adults have had time to gain job experience and/or further their education, many are in need of special assistance to assemble a business idea.

Stage 5 – *Growth*. Often, business owners do not seek help until it is almost too late. A series of continuing seminars or support groups can assist the entrepreneur in recognizing potential problems and how to deal with them in a thorough and timely manner.

Source: Consortium for Entrepreneurship Education, Columbus, OH, http://www.entre-ed.org.

Higher enterprise education in Universities and Higher Education Institutions should be considered a noble activity, worthy of the highest endeavors. An entrepreneur needs to be good at many things. Interdisciplinary academic institutions that offer integrative links between disciplines help people to set up their own businesses. Yet, a cross-disciplinary approach is not the sole characteristic that qualifies entrepreneurship education. Baumol (2004) has established a difference between "education for innovative ideas" and "education for mastery of received knowledge," with the former constituting a truly creative environment for inventive entrepreneurs.

Exhibit 77: Educational Preparation for Innovative Ideas and for Mastery of Received Knowledge

Baumol draws a dividing line between founders of business firms who are "inventive entrepreneurs" responsible for novel innovations (breakthroughs) and those who are "incremental innovators" engaged in cumulative incremental improvements.

The educational preparation of the incremental innovator leads to mastery of the already available paths of scientific knowledge and methods. The inventive entrepreneur needs an unorthodox approach to education that favors the free-wheeling exercise of the imagination. In Baumol's words, "We know little about training for the critical task of breakthrough innovation."

Source: Baumol 2004.

The world is the real community of and for entrepreneurship. The exposure to the international world of entrepreneurship education by means of educational exchange programs, business projects, and student trips leads to the sharing of ideas about business and making business between different cultures and countries. Therefore, entrepreneurship educators must expose their students to the great experience of international relationships.

Building fluid, productive multicountry networks is the new requirement for cross-boundary knowledge sharing and collaboration. Economic, organizational, and interpersonal relationships among students of different countries and cultures are likely to result in international entrepreneurship.

Educational institutions should help entrepreneurial students meet and learn reciprocally from their colleagues worldwide. Mobility within international learning networks shapes informal circles of exchange that foster a cross-border entrepreneurial behavior. Therefore, an effective entrepreneurial strategy embraces teaching entrepreneurship and the setting up of learning and support networks for future entrepreneurs and start-ups. This can also include project-based exchanges for students, who can move abroad to study, take a job experience in another country, and later return home to establish a new firm that can be born global or to conduct business across national borders – thanks to the founder's international relationships. This form of brain circular migration is the dynamo powering a country to international preeminence in the new age of the knowledge-intensive entrepreneurial economy.

Exhibit 78: New Educational Contexts for Building Relationships Among Like-Minded Individuals of Different Countries and Cultures

In the context of the Global Village Program at the Iacocca Institute, its Director, Brandt (2003), has observed: "Young entrepreneurs work with other young entrepreneurs in other countries because these people were seven-day-a-week classmates together in the six week Global Village Program, and lived together and became like family. They know how each other thinks, and most importantly, they trust each other with their business, and together they can shorten the time cycle for entrepreneurial business." The big issue here is trust and understanding, as shown by the two following examples:

A president of a small generator factory in Italy, now travels the Middle East with a fellow Global Village intern who is from Iraq, acts as his agent, but lives in Sweden and has many contacts in the Middle East. A young entrepreneur with a successful transportation business in Mexico visits his fellow Global Villager from Argentina who also has his own water bottling business, and together they investigate and invest in bringing an upscale leather business back to Mexico.

The University of Oslo has initiated a program called the Norwegian School of Entrepreneurship sponsored by the Norwegian government. The program consists of internships abroad with evening entrepreneurship classes at local universities every summer. Students come from seven different universities in Norway with a variety of backgrounds from a minimum of 3 years in higher education to doctoral students and people with industry experience.

In the case of the International Entrepreneurship Academy (http://www.intentac.org) at the Jönköping University in Sweden, student groups are turned into an international learning network of cross-functional expertise. A worldwide range of educational and business partners endows the Academy's programs with the image of a collegiate community of knowledge practice.

Notes

[1] Serendip is an ancient name of Ceylon. It seems that the word "serendipity" was coined by Horace Walpole, who based it on the title of this fairy story (The Wordsworth Dictionary of Phrase and Fable: 981).

[2] The composite index of cultural integration is the sum of the rankings of three individual integration measures: Kluver and Fu's index of cultural trade and Kearney/Foreign Policy's indexes of technological connectivity and international travel (Kluver and Fu 2004).

The measure of creative occupations includes professionals, artists, musicians, scientists, economists, architects, engineers, managers, and other workers whose jobs deal with creative, conceptual tasks (Florida and Tinagli 2004).

Chapter 11
Entrepreneurial and Corporate Universities

He that will not apply new remedies must expect new evils; for time is the greatest innovator.

(Francis Bacon)

Better-educated individuals raise the potential to start new businesses. Investment in tertiary education ought to be diverted toward innovative educational institutions that allow people to acquire skills they need to recognize and pursue business opportunities. The entrepreneurial universities are the most advanced forms of educational institutions embarking upon a new wave of teaching and learning methodologies in the field of entrepreneurship. Besides, these institutions harvest, in the marketplace, the fruits of university research, capitalize on business development thanks to the know how of professors, researchers, graduates and students, and provokes new company formation in the knowledge-based industries.

According to David Blunkett, the United Kingdom's former Secretary of State for Education, "In the knowledge economy, entrepreneurial universities will be as important as entrepreneurial businesses." In fact, to address the phenomenon of entrepreneurship and accelerate its pace, the most dynamic economies are producing innovative types of social and business models for advanced education.

Traditional University

Universities traditionally enjoy a sense of elevated privilege as prime creators and diffusers of knowledge. Their organisation is based on broad coverage, stability, and resistance to change (Martin 2003). Scientific progress advances through a mixture of tacit and widely applauded steps, based on the established modes and frameworks within the particular relevant branch of the scientific community. The classic system for quality control, the peer review, has been designed for the purpose of producing the best possible knowledge while guiding learning processes based on the norms of science.

T. Andersson et al., *Knowledge-Driven Entrepreneurship*, Innovation, Technology, and Knowledge Management 102, DOI 10.1007/978-1-4419-1188-9_11,

The limits of the traditional and fairly uniform learning system of universities are becoming apparent, however, and in everyday practice gives rise to dilemmas how to handle conflicting demands. As the mechanisms underpinning widely accepted, trustworthy knowledge are subjected to new and varied impulses, the traditional monopoly position of the university is undermined (Beck 2003). In consequence, mainstream policy reforms expose science institutions to more competition, and aim to spur exchange of ideas on issues of significance to the national or regional innovation system.

While still the prime authority in knowledge creation, the university now meets with an intensifying and partly conflicting combination of demands, from accreditation agencies, markets, policy makers, and internal stakeholders. Some of these push for renewal and specialization, others rather for conservatism and streamlining. On balance, although there is too much herd behavior in learning as well as in industrial policies, with many running for accessing the same research fund or supporting the same market opportunity. There is a deficit of new investment opportunities generated by innovation, however, especially in China despite its rapidly increased competitiveness in manufacturing.

Entrepreneurial University

The entrepreneurial university can be defined as a fluid network of inter-linked agents performing specific activities and drawing strength and vitality from one another (Exhibit 79). They can also be "unbundled" or "disaggregated" from one another. This means that the entrepreneurial university is a self-organisation – that is, "a spontaneous formation of interest groups and coalition around specific issues, communication about those issues, cooperation and the formation of consensus on and commitment to a response to those issues." The organisation embraces the attribute of an adaptive system whose agents change rules of conduct as the system evolves (Stacey 1996: 333).

The self-organized network allows the entrepreneurial university to identify both the different markets or clients that they are serving and the different ways of connecting with and contributing to those markets or clients.

Information and communication technologies are important adjuncts to the network as they contribute to cut down coordination costs. But technology is neither a substitute for campus life nor for the entrepreneurial university as a community of scholars (including students and staff).[1] Indeed, the university ecology produces a collegiate effect between current and past students. It is expected that past students will continue to take part in the activities of the network by means of "learning contracts" signed by themselves as individuals or by their employers. A further vehicle for keeping them all in touch with the evolution of activities is the ex-alumni committee.

This type of university is the natural partner for companies and people capable of recognizing the potential of that market. Large corporations have already tackled the new business, investing in those education ventures called "corporate universities." Ideas and plans conceived for establishing entrepreneurial universities nowadays bring together even small and medium-sized companies from diverse sectors to broaden their understanding of the entrepreneurial economy.

Exhibit 79: Agents of the Entrepreneurial University

Degree granting bodies (DGBs): small and flexible administrative bodies which deliver degrees and set degree requirements and core courses. One DGB can take on as many or as few students and faculty as thought viable. Their main activities are the following:

Student orientation: competence oriented rather than diploma-oriented.

- Certificates and degrees
- Noncredit programmes
- Caree counseling

Faculty companies, which are independent contractors looking for DGBs to sanction their teaching. They might find more than one DGB to do this. Their main activities are the following:

- Lecturing and researching
- Mentoring and tutoring
- Academic counseling

Facility companies, which make available labs, equipment, libraries, and classrooms. Their clients might be several DGBs.

Incubator companies, engaged in nurturing start-ups from graduates and researchers in cooperation with strategic industrial partners and seed capitalists around the world.

Source: Adapted from Seely Brown and Duguid (1996)

Corporate University

Entrepreneurial universities are embedded in the entrepreneurial economy; corporate universities, in the managed economy. In the educational landscape, the former emerge from the activity of knowledge pools in the area of the entrepreneurial economy, while big companies have been creating the latter with the mission to offer in-house a continual knowledge and learning cycle for their employees. This can be viewed as the growing recognition by many large organizations that, on the one hand, the power of knowledge and learning is a fundamental driver of strategic changes, while on the other hand, world-class corporate universities can have a big impact on the exercise of that power (Prince and Beaver 2002).

It is important to note that the term "corporate university" embraces a diverse range of organisations with and without the word university in their title. A neutral definition is that of "an on-site learning initiative integrating credit and non-credit programs that are linked to organisational missions and goals" (Wells and Barley 1998). There are corporate universities that perform training functions and others acting as agent "for the management of change through [their] key role in facilitating knowledge management practices and operating the organisation's learning agenda." The latter aspire to the role of world-class corporate universities (Prince and Beaver 2001: 21).

Corporate education initiatives (see in Exhibit 80 the Infosys's Corporate Learning Initiative) strive to align corporate learning to business strategies, develop strategic learning alliances with external providers, create a learning environment through technology, develop and implement innovative marketing and branding techniques.

Throughout its long life cycle, the academic community claimed to have a monopoly on knowledge. But in the meantime, the business community has realized that, from a historical standpoint, the old-fashioned universities and schools were failing to provide industry with the skills it needs, and, from a business view, that higher and further education is a rising market, with even more demanding customers since ICT infostructures became available.

A renowned institution such as the Massachusetts Institute of Technology has recognized that the relationship between university and industry should not be based on the view that the university is the central source of knowledge, but on the conviction that the new knowledge and discoveries occur throughout society, and that the movement is never unidirectional.

In "A Survey of Universities" The Economist, October 4, 1997, suggested that the fashion for companies to establish their own universities "symbolises... the growing elision between the presumed jobs of the university (basic research, general education) and that of the company (applied research, job-specific training). McDonald, Motorola, Sun, Microsystems, Hewlett-Packard, Unipart, British Aerospace are only some of a many initiatives large companies are devoting to business education through corporate universities."

Yet corporate universities do not underrate the role in both education and creation of new knowledge played by the academic institutions and, therefore, they are interested in complementary forms of evolution with them. A growing number of corporate universities seems keen to develop complementary relationships with academia (Exhibit 81). Partnerships with accredited higher education institutions have been developing mostly in business administration, computer science, engineering, and finance.

Exhibit 80: The Infosys's Leadership Institute

"Infosys Technologies Ltd, a Bangalore-based company, is a world leader in consulting and information technology services.

"At the cornerstone of Infosys' Corporate Learning initiative is the Infosys Leadership Institute located in Mysore. The Leadership Institute was established to help manage Infosys' growth, prepare Infosys employees ('Infoscions') to face the complexities of the rapidly changing marketplace and bring about a paradigm shift in the work culture by instilling leadership qualities."

"Other components of the Corporate Learning Initiative are the Education & Research and Learning & Development departments, which, along with the Leadership Institute, enable the present and future leaders of Infosys to achieve the company's vision and strategy. Infosys exemplifies best practices in corporate learning, incorporating the organization's strategic goals into the learning infrastructure, and creating a learning environment that is aligned to and improves business performance. Infosys Corporate Learning leverages talent and leadership, maximizing through learning, value to their stakeholders – customers, community, board of directors, employees" (official praise from the Corporate University Xchange Excellence Award for 2002 received by Infosys Technologies Ltd).

Source: Infosys Press Release, June 7, 2002

Exhibit 81: Complementary Relationships Between Corporate Universities and Academia

"Our [corporate university] role is to be a bridge between the academic world and the business world. It is more and more important for companies to get a fresh infusion of knowledge to manage the permanent transfer of academic knowledge into the company" (Michael Heuser, head of Lufthansa School on Business in Frankfurt).

"The greatest asset of corporate universities is the daily access they have to senior leadership that allows them to align education programmes with the company's shifting strategic goals... Some corporations have joined with top business schools to create blended corporate degree programmes that offer the best of both world: a high-quality, accredited MBA programme and customised course projects that let employees do real work, not homework" (Thomas Moore, Dean of the executive education at Babson College).

Source: Business Education, Financial Times Survey, March 25, 2002.

Cultural Roots of the Entrepreneurial University

The twelfth-century term for a university was *studium* – a community of learning for students from all parts. The university of the Middle Age was a student-run institution. Students were the main protagonists in their own knowledge development. Consumer-students collected fees, paid salaries, issued the working rules, determined the content of the curriculum, and decided how much time had to be spent on each topic.

Contracts between students and professors were based on the criteria of outcome measurement rather than on attendance. Committees of students monitored the teachers' behavior. The idea behind these rules was that the students were serious about learning. The professors did not work for a board of trustees, as they do in modern universities, but for the students. If they were not meeting the students' needs, they were dismissed.

Throughout the Middle Ages, international mobility and exchanges of teachers and scholars were a common future of those collegiate societies. Learning was not separated from learners by artificial borders. The university community and the handicraft economy were interwoven, each having much to learn from the another. A transnational and collaborative context favored entrepreneurship, and innovative businesses were created that replaced traditional forms of arts and crafts.

Entrepreneurial universities are today's version of the medieval collegiate society. They may be seen as "academic impresarios" whose mission is to achieve a brand identity (Exhibit 82) by bringing together the best content specialists (outstanding practice-oriented academics and theory-oriented practitioners), junior teachers acting as "educational consultants," and students (Hague 1991). The integration of thinkers from industrial and consultant backgrounds with their academic counterparts strengthens the quality of educational programmes. Each partner adds value to the network, but the real value of it is greater than the sum of the individual parts. What makes the difference is a synergistic collaborative process involving people with complementary competencies, which results in a symbiotic learning network (Amidon 2001).

Exhibit 82: Designing a Brand Identity Policy

Brand background
Segmented into three wide categories – starting with the young at high school level, those that aspire to learn more at a tertiary level, and those who are already in the workforce.

Brand attributes

- Professional Focused – bringing together discipline and knowledge, in a framework that is focused on professional development within a global network or challenging individuals.

(continued)

Exhibit 82 (continued)

- Alignment – clear and concise way to maximize business potential. It is proactive and creative in approach, and fresh and customised in its outcome, through a collaborative and dynamic network.
- Momentum – building the continuous need for innovation. A pragmatic environment enabling students to compete in the global marketplace.
- Inspiration – a revolution in new thinking that will shape individuals and institutions to build and motivation cultural change toward entrepreneurship.

Hero product brands

- Corporate intrapreneurs
- Academic entrepreneurs
- Young entrepreneurs

Source: Enterprise IG, Brand & Identity Consultants, 2002

A Worldwide Network

An entrepreneurial university fosters interaction and networking in the same way that firms are expected to do, and in the same way that inventors such as Watt and Edison did, with scientists, economists, financiers, and other stakeholders and experts in the community.

The brand-owning entrepreneurial university operates in close cooperation with an external worldwide network of alliance partners (academic institutions, business firms, e-technology enablers, tailored groups of consumer-students, and so on) with whom it forms a value-added community (VAC) (Means and Schneider 2000). The VAC experiments with innovative forms of education in and for entrepreneurship, attracting a new generation of students – the *clerici vagantes* (wandering scholars) of modern times – who leave their own countries with the intention of being away for a couple of years or more. Usually, they return home having already implemented their entrepreneurial projects with the VAC's educational institutions and companies abroad where they have studied.

The university cities of the Middle Age used to harbor for a while students from other communities. Each of them played to his or her strengths, rather than ape the host university city. Along the route the *clerici vagantes* were pollinators of new ideas and projects that made the university cities wealthy. Nowadays a circuit of native missionaires like those in the Middle Age is a fertile ground for "glocal" communities, where the local dimension turns into a local and global dimension. One of the most striking examples of the formation of glocal communities is the circuit of students that links Bangalore, Hyderabad, Mumbai, Shanghai, and Beijing with London, Boston, and California (Fig. 18). Twinning entrepreneurial

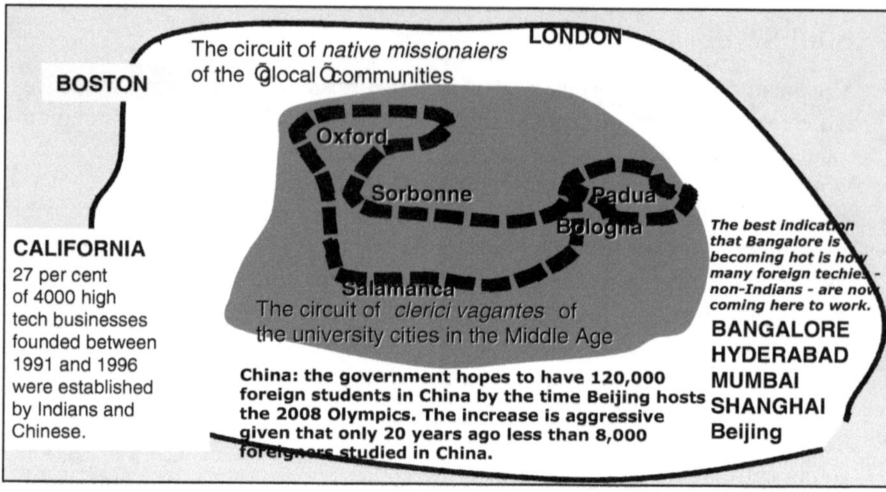

Fig. 18 Student mobility in the Middle Age and in the decade 1996–2006 "glocal" dimension

projects cultivated at the university sites within the circuit open up the door to successful entrepreneurial-friendly environments that, in turn, foster new and emerging high-growth business communities.

Open boundaries, education without borders, new connections, both physical and virtual journeys into other places and disciplines: all these are ingredients that foster new ideas. Thanks to mobility within the network, informal circles of exchange take shapes that are sources of creativity and cross-fertilization of ideas.

Entrepreneurial universities, then, design and manage global networks, onsite and online, which are created and developed by means of worldwide alliances with learning partners and business organizations that link student-centered learning to on-the-job activities. In such a cooperative environment, participants can cultivate new business ideas and turn them into commercial realities.

Participants can move from one learning location to another, and, in each location, the diversity and ethnic mix of both the student population and the faculty members play an important part in reducing the risk of a brain drain from developing countries and regions.

What Is the Policy Response?

There is not a single answer how to respond. However, there is a genuine need of allowing for diversity and pluralism in university governance (Andersson 2008). Across the board, university management needs to act strategically, while learning how thereby to boost rather than undercut creativity and initiative from below. If they are to remain the prime knowledge producers, and serve as a cradle for those innovation that has potential to bring what is truly new, universities must be able to test and improve their own ways of governance, based on strive to cherish the strengths and opportunities that pertain to the specific institution.

The ability of universities to do just that is, in practice, restricted due to a combination of regulation, funding principles, and tradition. As an illustration of the implications, note that *Jönköping University*, a young independent foundation being one of only three universities in Sweden not subjected to the restrictive conditions of a public authority, enjoys the *atypical* advantage to run board meetings in English (thus not limited to having Swedish-speaking board members), starts its own holding companies over night not having to wait for years for the uncertain outcome of government examination, controls its own premises, runs its own endowment, organizes 850 small and medium-sized companies as mentors for all its students in business and engineering (the last not the result of legal advantage, but out of sense of identity), and so forth.

Policies should be framed in recognition of the need for multiple and inherently diverse and complementary institutions engaged in higher education, research, and innovation. Each university must then be able to foster its specific governance model, capable of serving as an instrument for day-to-day learning as well as for setting long-term direction, applying across the range of university functions, including education, research, innovative and path-breaking relations with wider society, whether in the form of big business, SMEs, hospitals, schools, and so forth.

The optimal combination of continuity, on the one hand, and that of preemptying and adjusting to changing conditions, on the other hand, is bound to vary between institutions. Reforms spanning regulation and funding, promoting student mobility, and pushing better information provision should aim to widen the scope for entrepreneurial universities so as to promote greater diversity through an organic process, and so as to ensure that increased *autonomy* goes together with increased *accountability*.

Note

[1] As for online courses, on the one hand, educational technology specialists worry about the quality of the teacher–student relationship when long-distance education increases. On the other hand, a widely held belief is that computer facilities better than classrooms help customize individually tailored curricula (Pescovitz 1996).

Chapter 12
Small Business and Entrepreneurial Growth Companies

There is no finer investment in any community than putting milk into babies.

(Winston Churchill in a radio broadcast to the British nation, March 21, 1943)

The quality of entrepreneurship evokes the difference between small business ventures and entrepreneurial growth companies.

The creation of entrepreneurship in terms of the quantity of new ventures does not automatically assure profitable changes. The success of the enterprise economy machine depends on the quality of its engine – that is, the process of allocating resources for the generation of productive and innovation-driven rather than rent-seeking undertakings. Rent-seeking activities promote forces that choose corporatism over enterprise, and are responsible for a blatant abuse of power. In contrast, productive enterprises constitute the army trained to fight on the ground of competition.

On the one hand, a small business venture tends to be independently owned and operated, not dominant in its field, and not engaged in new marketing or innovative practices. On the other hand, the principal goals of an entrepreneurial venture are profitability and growth, and the business is characterized by innovative strategic practices.

A small business owner is an individual who establishes and manages a business for the principal purpose of furthering personal goals. The business tends to be the primary source of income and will consume the majority of one's time and resources. The owner perceives the business as an extension of his or her personality, intricately bound with family needs and desires. By contrast, entrepreneurs such as those typified in Chap. 10 establish and manage a business for the principal purpose of profit and growth.

T. Andersson et al., *Knowledge-Driven Entrepreneurship*, Innovation, Technology, and Knowledge Management 102, DOI 10.1007/978-1-4419-1188-9_12,

Small Business Ventures

Small businesses have been depicted as the epitome of entrepreneurship, and entrepreneurial communities have been defined as those communities in which small business founders have formed a free, self-catalyzing network whose members act like a flock of birds, aligning with one another and flying in formation, as shown in Fig. 19. Indeed, in these communities (often also referred to as "industry clusters") entrepreneurial moves can be identified that resemble the motion of a flock of birds flying, as described in the model created by Craig Reynolds and revived by Gary W. Flake (Flake 1999).

In Fig. 19, the bird framed in black represents a new, small business founder. His or her movements are articulated in four sequential steps:

- *Step 1* is the *differentiation* movement. As each bird avoids flying too close to another in the flock to reduce the chance of a mid-air collision, so the new founder attempts to make his or her business distinct from the others in the cluster.
- *Step 2* is concerned with *copying or imitation*. After a successful start-up phase, the new company's founder conforms to the group's convoy mentality. Replicating the movements of the neighbors, the entrepreneur now follows the general direction of the group. The prevailing pattern of competition is that of the imitative competition close to perfect competition, which requires a large number of firms in an industry. Overall, the outcome of the small business is a bounded performance in terms of limited growth prospects, albeit occurring with a lasting profitability – which is, however, judged as a successful accomplishment or even a "cause for celebration."
- A *move toward the center* constitutes *Step 3*. Protection reinforces the imitation move. A defensive perspective incites the small business founder to take a position at the center of the perceived cluster/flock where there is the least

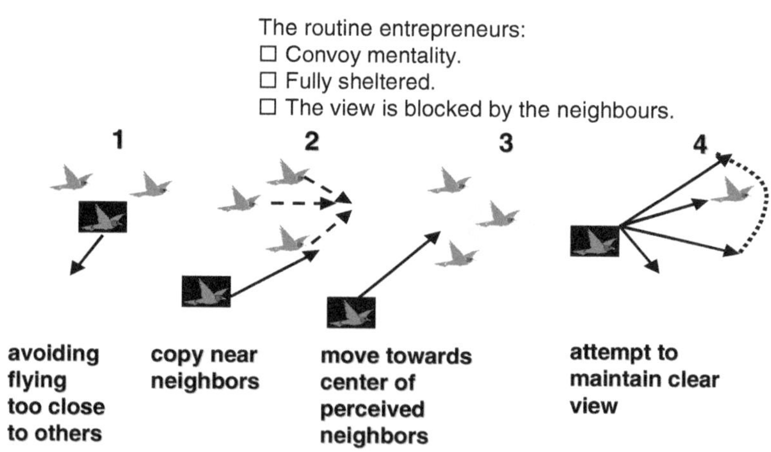

Fig. 19 Small business-type model of entrepreneurial motion
Source: Adapted from Flake (1999)

exposure to external threats. However, in taking this shelter, his or her view is obstructed by neighbors. Deprived of a range of vision, the entrepreneur will lose the capacity for differentiation that requires lateral movements. This prompts the small business founder to move to:

- *Step 4* in which he or she attempts to maintain a clear *view* of future prospects.

These movements explain why most businesses start and remain small, even after many years of trading. For them, the growth period never occurs, or else it comes too slowly and too late, sometimes even as a precursor to the death of the business.

There is a trade off between pursuing a comfortable state of equilibrium, which maintains the independence of the founder's small company and provides economic support for his or her family, and enhancing the founder's entrepreneurial capacity and entrepreneurial opportunities. Dumping down both of them is the price paid for self-preservation and maintaining this unimaginative approach and the status quo. Therefore, the small business does not transform into growth company. Rather than being focused on growth, it fights for financial self-sufficiency.

As the Global Entrepreneurship Monitor (GEM) has remarked, the owner-manager of a small business is inclined to be financially and managerially self-sufficient, having little or no recourse to external resources. A self-imposed limit to growth discourages the development of growth-oriented firms.

GEM has founded that "It is much more common for a firm to be started than for a firm to be entrepreneurial…and [that] the overwhelming majority (95%) of existing firms continue to replicate existing activity and are not very entrepreneurial" (Babson 2003).

Entrepreneurial Growth Companies

Fresh competition in the free market economy and the narrowing down of international borders significantly influence new company formation. Ample opportunities for creativity and innovation are driving the move toward the formation of new ventures whose founders are focused more on growth and less on self-sufficiency. From the start, they are ventures that enter into a fast and high-growth phase – the so-called entrepreneurial growth companies (EGCs).

EGCs do not necessarily are the ones that first make a great discovery. Findings reported by the US National Commission on Entrepreneurship show that often times "they tend to make smaller innovations in products or processes and then perform exceedingly well. As companies fund research and make discoveries, they tend to develop breakthrough technologies". Neither should we take it for granted that ECGs founders are expert in that particular field in which the company is embedded. The same report points out that "entrepreneurs with little expertise in their fields have started many of the most successful entrepreneurial companies."

For high-growth business, the following steps are discernible in Fig. 20:

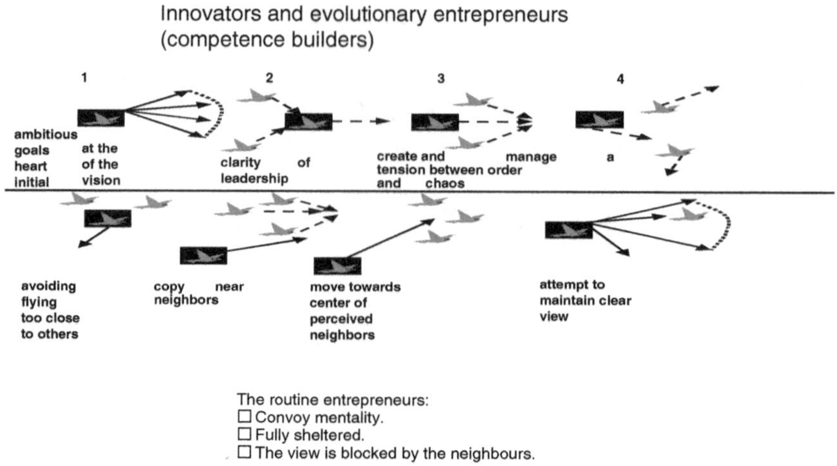

Fig. 20 Entrepreneurial growth company-type model of entrepreneurial motion
Source: Adapted from Flake (1999)

- *Step 1. Audacity*. "To win a lottery you need to buy a ticket." Entrepreneurship is a small business "with a lottery ticket attached" – reports the National Commission on Entrepreneurship, quoting Amar Bhidé, who has written seminal papers on the origin and evolution of EGCs. EGC founders have bold and ambitious aims. Identifying business opportunities different from those already available in the market is not enough. Compared with small business founders, their vision is much more audacious. EGC founders aspire to create companies whose course of growth is well defined from the beginning – often at an annual rate of 15–25%. Therefore, they capitalize on opportunities convertible into marketable products and services that offer potential productivity gains higher than those selected by small business founders (US National Commission on Entrepreneurship, 2001b). The prevailing pattern of competition is that of the innovative competition resembling a race in which the winner takes all not because he obtains a decisive advantage, being, eventually, the first-mover in the game, but because he can deliver a plainly superior product (Liebowitz 2002).
- *Step 2. Leadership rather than experience in their fields marks a difference*. In pursuing their audacious goals, EGC founders cannot act as followers who imitate close neighbors. To run fast and win, they have to be forerunners. This does not necessarily mean that they have to neither be experienced experts in their fields nor that their companies must be based on breakthrough technologies. The clarity of leadership is their distinctiveness.

In contrast to small business founders, who are catalyzed by a cluster, entrepreneurs in EGCs head for new cluster formation and are thus catalysts of small business growth and development. They are driven by the motivation to share, whereby collaborating with other firms instead of going it alone becomes the essence of their business – as found in a survey of America's fastest growing companies from Pricewaterhouse Coopers, which estimates that collaboration has generated bottom line benefits, accounting for nearly a quarter of current revenue.

- *Steps 3 and 4: Surfing the edge of chaos.* EGC founders create and manage tension between order and chaos. The success of growth companies is dependent on a ceaseless process of revitalization that allows for fluid, open boundaries, and new connections. Between the equilibrium of an orderly cluster and the disequilibrium caused by a fundamental disturbance that explodes in disruption, surfing the edge of chaos maybe uncomfortable but necessary for fast-growing new businesses.

Fast-growing start-ups in every economic sector from high tech to manufacturing are the main engines of wealth and job creation. These are companies that move technology into society ("technological start-ups"), move processes from one part of the world to another ("geographical start-ups") and change ambiance or method of delivery ("sociological start-ups"). They account for two-thirds of the differences in economic growth rates among industrialized nations. According to the PricewaterhouseCoopers's survey of America's fastest growing companies, these types of companies have contributed to the creation of two-thirds of net new jobs and have been responsible for more than two-thirds of the innovation in the US economy (and 95% of all radical innovations since World War II).

Exhibit 83: The Importance of New and Emerging Entrepreneurial Growth Companies: The "Two-Thirds" Change Agents

Small entrepreneurial companies cover a small proportion (5–15%) of all US businesses. Yet data collected over the last 20 years show that these companies:

- Create two-thirds of net new jobs.
- Are responsible for more than two-thirds of the innovation in the economy (and 95% of all radical innovations created since World War II).
- Account for two-thirds of the differences in economic growth rates among industrialized nations.
- Regularly collaborate with other companies. Overall, 56% of fast-growth companies worked collaboratively during the past 3 years. Twenty-nine percent of the collaborations focused on new product development, while 37% sought to improve existing products. These partnerships were not one-time affairs.

Source: US National Commission on Entrepreneurship 2002c, d; PricewaterhouseCoopers's survey of America's fastest growing companies (http://www.barometersurveys.com/)

Chapter 13
Native and International Entrepreneurship

Why join the long queue, when you can join the short one?
(Asian proverb)

A new anatomy of the entrepreneurial body is emerging with increased mobility of people who travel, visit, study, or work without restraint outside their native countries. Moreover, the Internet-mediated economy facilitates personal ties and continued access to one's home culture. Physical proximity is, therefore, no longer the most important factor in discouraging networking on a world scale.

Mobility across borders and novel scientific-technological capabilities are driving a transition from the traditional form of entrepreneurship (we call it "native entrepreneurship," to typify the locally oriented and static behavior of the entrepreneurial context) organized along lines of geographical proximity and cultural identity to a new form called international entrepreneurship, which is one of the most visible manifestations of a process of cultural integration that happens through a global spread of ideas.

Exhibit 84: Understanding the Nature of International Entrepreneurship

Local relevance and global competitiveness define the nature of competition in today's product and service markets. Locally oriented entrepreneurs must be attentive to competition from international players. Concerns arise over the long-held tenet that start-ups, such as existing small businesses limited to domestic markets, could reach out beyond their turf through incremental, step-by-step patterns of internationalization. They have to contend with the physical distance and the level of cultural difference from their target markets, and often they are forced to cede their niche markets to global competitors who are able to grow the small market into a much larger one.

Unlike the past when entrepreneurs could independently determine their critical tasks on the basis of physical resources that used to be local, the years from the early 1990s to the present have seen an ever-rising number of entrepreneurs capitalizing on intangible knowledge assets whose sources are international. In fact, intellectual capital is basically an issue of context that requires a world of relationships, alliances, fluidity, and cross-border networking.

T. Andersson et al., *Knowledge-Driven Entrepreneurship*, Innovation, Technology, and Knowledge Management 102, DOI 10.1007/978-1-4419-1188-9_13,

Native entrepreneurship resembles an island whose borders are dictated by the natural barriers of its physical space. What is more, the proximity effect creates other types of barriers – those raised by family favoritism, crony capitalism, ethnic, racial or religious factionalism, which have oftentimes prevented native new ventures from succeeding.

International entrepreneurship is embedded in a borderless context where innovations are spawned by science-based revolutions and turned into networks of new products and services that are simultaneously launched all over the world. Natural barriers are no longer borders to communication and trade.

International entrepreneurship owns a concerted nature that comes to life via distributed networks formed among prospective and current entrepreneurs using the technologies of globalization. Their experiences and practices of transnational, cross-border and cross-cultural relationships make possible new venture creation with an international focus from the start. The concerted nature of international entrepreneurship has significant and positive economic impacts. It serves to reduce barriers to and costs of communication between entrepreneurs of different countries and cultures. By enabling more effective and constructive combinations of complementary skills, concerted entrepreneurship enables positive synergy effects, or "strategic complementarities," from the efforts of different actors. Although marked by disparate interests and backgrounds, these entrepreneurs become more able to address shared problem or processes in a consistent manner, to create international firms, and to provide spillover effects by lowering search costs and raising skill levels among other would-be international entrepreneurs, who can benefit from the experiences and practices of previous explorers in this rapidly evolving field of business.

The most significant barriers are symbolized by differences in language, customs, legal systems, religions, and, notably, trust building beyond the family context and cultural contiguity. Because international start-ups can become frustrated by these cultural barriers, forming and strengthening shared cultures would lend a decisive contribution to their lowering and make it more likely for any person anywhere to cooperate with any other person elsewhere.

The mental models nascent entrepreneurs developed in a given country shape their perceptions about how market players and institutions behave in another country with a different socio-cultural and linguistic environment. International entrepreneurship alters those subjective representations. Once enterprising founders of business firms in country A form contacts and discuss ideas with their peers in country B, the former redefine their mental models to explain and interpret markets in B and vice versa. Thus, international start-ups resulting from a team of founders from both countries shape choices, define ways of doing things, develop communication practises, and make market connections that traditional start-ups could not articulate as well.

The benefit brought about by international entrepreneurship is twofold: on the one hand, the market of the international start-up does not appear too small initially; on the other hand, the capability of local adaptation to the target market leads to a reduction in the economic cost of business mistakes.

Exhibit 85: International Entrepreneurship

International entrepreneurship is the discovery and exploitation of business opportunities across national borders. International entrepreneurship adds a substantial new dimension to the entrepreneurial economy, reflecting environmental experiences that from the very beginning move new ventures into an international marketplace.

"A new form of entrepreneurship is developing. Instead of focusing just on one country, today's innovative start-ups are increasingly looking globally for ideas, funding, people, and markets. This is particularly true for new companies in Latin America, Western Europe, and Asia. It is also true for many new companies in the United States" (Halperin 2001).

International entrepreneurship embraces entrepreneurial activities that cross national borders. New venture firms engaged in internationalisation from inception have been referred to as international or born global start-ups. Entrepreneurial behavior that extends across national borders requires local market knowledge. The management of knowledge is particularly challenging for the discovery and exploitation of international business opportunities that have to deal with differences in cultures, languages, corporate governance systems, public policy frames, etc. (McDougall and Oviatt 2003).

An innovative form of international start-ups is that which emerges from networks of would-be founders from multiple countries: for example, groups of students enrolled in entrepreneurship education programs organized by universities and business schools of different countries. A network of relationships between them could have more influence on the creation of international new ventures than does their physical distance.

International Start-Ups

International start-ups are emerging as an innovative breakaway pattern of entrepreneurial activity. Instrumental in setting the trend for international start-ups has been the dramatic shift in entrepreneurial environment from a local to a trans-national focus. The new pattern of entrepreneurial activity makes traceable a divide between the "one man show" – the solo entrepreneur surrounded by rigid firewalls – and a range of complementary entrepreneurs for complementary innovations (Bhidé 2000). This new species of entrepreneur creates international and stateless companies, which contribute to an increase in the variety of enterprises in the start-up population ecology.

By mirroring an environment where borders separate culturally divergent and even small markets, the normal process of entrepreneurship figures on the solo inventor-entrepreneur whose control upon local/national physical capital assets is a distinctive piece of his operating policy, whereby he translates a business idea into action. He makes his decisions looking first at the domestic market. In most cases, it is not within a few short years that this local venture focused on local goods and services starts to target foreign markets where, to be effective, it opens sales offices inasmuch as decisions of splitting the company in two or more locations are likely to pose extreme organisational challenges.

The inbuilt advantage of the international start-up over the national one is that the former can spot a larger market, for each member of the whole group of founders has an understanding of the native market from the ground up – a grassroots rather than a bird's-eye view. Each of the founders, knowing the market they are preparing to be in, can, therefore, develop a larger portfolio of profitable opportunities and fulfill the potential to match the needs of consumers in that specific market.

Building cross-border and across cultural and ethnic boundary start-ups is a process that needs to be nurtured to a larger extent in the atmosphere of low context communities (see Chap. 3). From this point of view, there is a twofold role that people mobility could play in shaping low context communities. In one respect, mobility in a physical sense stimulates face-to-face communication. On another, mobility in a virtual sense, which takes advantage from new open space technologies, is the wellspring of online knowledge communities. What looks promising for the creation of those start-ups is the coordination of and reciprocity between face and non-face-based communities.

A study on the internationalization of high tech start-ups at or near the inception found that "that the principal factors influencing the early internationalization of international start-ups are the international vision of the founders, their desire to be international market leaders, the identification of specific international opportunities, and the possession of international contacts and sales leads" (Johnson 2004).

With the purpose of expanding the pie in a context of win–win relationships, founders of international start-ups draw resources from and sell their goods in multiple countries from the very early stage of their development (McDougall and Oviatt 2003).

Among international start-ups, there are those new ventures whose scope extends well beyond their globally dispersed mode to an organization without a clear national identity (variously defined as "stateless," "global born," "transnational") that thrives on the diversity of its cross-country founders. There needs to be a high level of trust between the stateless start-ups' founders – which allows them to be split between the different locations with the twofold role of developing high trust relationships between the cross-border parties and teams, and operating the globally dispersed units as if they were one (Halperin 2001).

A clear mission to go cross-border from the inception phase portends that founders of international and stateless start-ups have to be focused not only on their own decisions but also on how those decisions affect other peers whose "tacit insider knowledge" could be a springboard for new business opportunities through collaboration. It is through a "revolving door" that they can access a global talent pool to build much-needed culturally diversified hybrid teams, in order to succeed.

Exhibit 86: Key Features of European Fast-Moving Stateless Start-Ups

- No particular national identity
- Trans-national focus
- International from the start
- Rapid international expansion for success
- Thriving on cultural diversity
- Intercultural settings
- Founders and other key persons are split between the different locations of the start-up to develop high trust relationships cross-site
- Building hybrid teams to succeed
- Assembling management teams from a global talent pool
- Educational experiences in business schools with highly diverse student bodies and international exchange programs
- Each team member speaks several languages

Source: adapted from Business Week – International Report on Stateless Start-ups, November 20, 2000

Exhibit 87: Internal Arrangements in the Institutional Structure for International Start-Ups

- Start-up units located in different countries operate as if they were one.
- All sites keep feeling that they are crucial to the success of the operation.
- High level of trust between the diverse sites contributes to high performance across distance.
- Founders have the skill and patience to pass on their knowledge to their cofounders.
- Personal interaction facilitates the transfer of tacit knowledge.
- Having a mix of locals and foreigners at each site of the start-up ensures communication.
- Ties between founders and nonfounders are as strong as within the founders' team.
- Common values are emphasized to avoid affective conflicts.
- All distributed teams are treated equally.
- The location of board meetings is alternated between the diverse sites.

Source: adapted from Halperin (2001)

Cultural Integration and Cultural Diversity

Globalization, openness, and integration call for cultural integration as a fertile ground on which shared cultures can be cultivated. "Cultural proxies" such as the cross-border movement of people and intellectual transfers can measure a country's level of cultural integration. Cultural trade, identified by exports and imports of books, academic journals, periodicals, magazines, and newspapers, is one form of intellectual transfer. Other components are technological connectivity (i.e., internet users, internet hosts, and secure servers) and international travel. Kluver and Fu's (2004) ranking of the most culturally integrated countries in terms of cultural trade, by measuring "each nation's exports and imports of books, periodicals, and newspapers," as well as indicators of technological connectivity and international travel, which are included among those indicators used for the construction of the annual A. T. Kearney/Foreign Policy Globalisation Index (2004), help direct the attention toward those countries that could more likely give a boost to the process of international entrepreneurship.

In Fig. 21, a composite index of cultural integration for 20 European countries is shown. This index is the sum of the rankings of three individual integration measures: Kluver and Fu's index of cultural trade and Kearney/Foreign Policy's

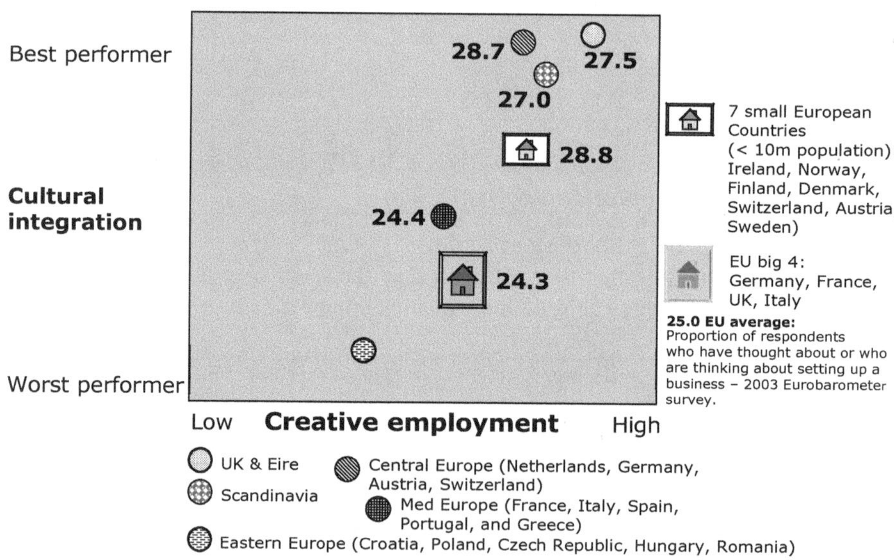

Fig. 21 Cultural integration, creative employment, and entrepreneurial tension in selected European regions.
Source: Kluver and Fu (2004); Florida and Tinagli (2004); A. T. Kearney/Foreign Policy Globalisation Index (www.foreignpolicy.com/wwwboard/g-index.php), EOS Gallup Europe – Flash Eurobarometer No. 146 <<Entrepreneurship>>, 10-23/09/03 – Report, p. 20

indexes of technological connectivity and international travel. The cultural integration index is axised against the Florida's index of creative occupations[1] as a percentage of total employment. Although a third index that measures international entrepreneurship is not available, a pattern emerges from the comparison between the level of cultural integration, the level of creative employment, and the entrepreneurial spirit or tension, the latter measured by using as a proxy the number of citizens in the EU countries who have thought about or who are thinking about setting up a business, according to the data provided by the Eurobarometer on entrepreneurship.

Figure 21 indicates that the level of cultural integration is positively correlated with the level of creative employment, and both are positively associated with entrepreneurial tension. In terms of size, small economies (<10 m population) are bettered suited to the requirement of a dynamic entrepreneurial economy in comparison with the Big 4 EU countries. Smaller economies exhibit higher levels of cultural integration, creative employment, and entrepreneurial tension. As for geographical location, overall the Nordic countries are better performers than those of South and Eastern Europe. There are three groups of countries at the top in the rankings of cultural integration and creative employment that also exhibit an entrepreneurial spirit above the EU average: UK and Eire, Central Europe and Scandinavia. The Mediterranean countries hold an intermediate position, whereas Eastern countries rank near the bottom.

It is worth noting that the pattern shown in Fig. 21 is consistent with the findings of the Lisbon Review, 2002–2003 (World Economic Forum 2002). This report points out that the three best performing EU countries as regards conditions for start-ups are Finland, UK, and Ireland, whereas the four worst performing countries are Greece, Italy, France, and Spain.

Exhibit 88: Globalization, Openness, Integration

Globalization refers to the extent to which trading relations of an economy have come to span the globe. It implies the diminished importance of geography to that trade.

Openness refers solely to the volume of trade in terms of the value of imports plus exports as a share of a nation's gross domestic product.

Integration implies increasing trading relations within a given economic area (e.g., European Union). In the EU's countries, the rate integration with one another far outweigh their rate of integration with non-EU economies.

Source: Colin Hay (2002), Tough questions for Europe, Financial Times, Friday August 2, 2002.

Interactions between would-be founders across national borders raise the issue of inescapable differences in culture and national character. For example, there are profound and critical differences with regard to attitudes toward corporate governance and funding patterns. Relying on diversity in cultural background to build start-up teams with no clear nationality entails a strong culture of innovation. Indeed, a common approach to business opportunities, a common language, and convergent beliefs no matter where founders and team members come from seem to be distinctive features of innovative, fast-moving international start-ups (Edmondson 2000).

The Lunar Society, a cross-border community of knowledge practice sketched in Part One, Chap. 4, is perhaps the most noticeable example of entrepreneurialism backed by bringing together persons of all distinctions. For their ability to do many different kinds of work successfully, most members of the Lunar Society might be called great entrepreneurial tinkers.

Thriving on cultural diversity for rapid growth, international start-ups gives fresh weight to the argument of innovation as the much-needed cultural key by which the unavoidable differences in national character are dimmed. The international market focus of fast-moving start-ups means that they approach new opportunities by means of a wider set of convergent beliefs.

Exhibit 89: Cultural Diversity: A "Plus' for Rapid Growth Start-Ups

There are different sources of cultural diversity. Imaginatik, a Boston, MA-based software and professional services company specializing in Innovation and Idea Management, has detected nine forms of diversity:

"Personal style (diversity of thinking)
Timing (different views of the future)
Gender (men and women have different needs and thought processes)
Ethnicity (race, religion, upbringing)
National culture
Age (every generation has its own way of thinking)
Work experiences
Life experiences
External viewpoints (customers, suppliers, partners, and distributors)"

In a research conducted at the Sloan Management School, the author submits, "Cultural differences can be used as a competitive advantage. Management should lift the taboo on discussions of cultural differences. Cultural stereotypes are a useful starting point for a learning process on cultural differences. Stereotypes can be used as a tool for understanding preconceptions and building cross-cultural teams. The cultural learning can become a unifying team experience."

Source: Imaginatik (2004); Halperin (2001)

Yet, cultures can still clash. Thus, borderless thinking, cultural curiosity, and cross cultural work of the international start-up have to be used as vehicles for

Exhibit 90: Cultural Diversity: The Lunar Society, 1765–1813

The Lunar Society was a remarkable grouping of people from varied backgrounds: mostly, provincial manufacturers, professional men and gifted amateurs who met every month in Birmingham on the Monday nearest the full moon from 1765 to 1813. They believed in entrepreneurialism to which they had much to contribute through their interest to the application of science to manufacturing, mining, transportation, education, medicine, and much else.

Their achievements were the outcome of a positive and proactive attitude toward cultural diversity. As one of the Society's member, Joseph Priestley – a minister of religion and amateur scientist who discovered oxygen, wrote: "We had nothing to do with the religious or political principles of each other. We were united by a common love of science, which we thought sufficient to bring together persons of all distinctions, Christians, Jews, Mohametans, and Heathens, Monarchists and Republicans."

Source: Uglow (2002: xiv).

exchanges of experiences with regard to the "identity" of the targeted regional markets. Start-ups whose operations are across borders enter into business environments that often show unique traits in terms of institutional framework, market structures, industrial and associational organisations, social relations, and cultural practices. In the light of the barriers raised by these distinctive characteristics, international start-ups need actors who are familiar with the social milieu and modus operandi of host business environments.

From the assumption that cultural diversity produces creative thought, it does not necessarily follow that new ideas automatically translate to each and every country. An international start-up can be the outcome of a business idea by one of the founders, say from U.S. Yet, his/her creative thought does not automatically match the market needs of the cofounders' countries, say from Europe and Asia. To fit the national market of each cofounder, that idea generation needs to be integrated by the "insider" cofounder. Thus, the odds of succeeding outside one's home market bears much relation to whether and to what extent the international start-up is endowed with a mixed background that covers both cultural diversity and regional identity. In addition, it has to be adept at tapping into a global talent pool to form a hybrid management team and enticing knowledgeable founding entrepreneurs from the targeted markets, who complement each other on each other's home turf.

Tapping into a Global Talent Pool: The Student Mobility Dimension of International Entrepreneurship

There is a growing area of consensus around the assumption that the international mobility of students is a vehicle for superior participation in the two types of entrepreneurship education: the camp that sets the stage for incremental improvements and the camp that sets in motion a process of thinking in unconventional ways that comes up with heterodox ideas and proceeds in unexplored directions.

According to the OECD "talent is disseminated most easily through the physical movement of people... the international mobility of students is crucial in deciding which countries are most likely to take advantage of new ideas." The international mobility of talented youngsters helps countries, regions, and territorial communities close their productivity gap vis-à-vis the most advanced economies, since it promotes entrepreneurship-led innovation. This reduces the risk of talent drains into the most advanced economies – a vast phenomenon that could provoke lasting economic repercussions in those countries, mainly the emerging ones, affected by the brain drain.

Open boundaries, education without borders, new connections, physical and virtual journeys into other places and disciplines: all these are ingredients that foster new ideas. The spirit of entrepreneurship is unconstrained by conventional cultural, political, and geographical barriers. Findings from studies of entrepreneurship in different cultural contexts indicate that there are common values shared by entrepreneurs despite the diversity of their cultural roots (Cox 1997).

The student mobility dimension evokes two societal breakthroughs that, respectively, the Phoenicians and medieval communities of scholars were made by intuition rather than through a laborious linear logical process, which was the style of innovation embraced by the ancient Greeks.

The Phoenician travels and displacements have spanned geographical barriers and shrunk the world of education. They generated interactive spaces for knowledge creation, knowledge dissemination, and knowledge sharing. The circulation of professors and students, and the resulting exchange of ideas in a climate of freedom, gave birth in medieval Europe to centers of higher education as guilds of wandering scholars (*clerici vagantes*). The first was the Studium of Bologna around AD 1088.

Both the Phoenicians and medieval scholars showed how the culture of transnationality produces "flexible citizenship...that induce[s] subjects to respond fluidly and opportunistically to changing political-economic conditions" (Karam 2001). That form of mobility and the resulting intellectual exchanges foreshadowed those processes of cultural integration, knowledge creation, and result-oriented innovation actions that unfolded all through the Renaissance movement. Brain circulation triggered off nine inventions that were developed during the Renaissance: Clocks, Gunpowder and Artillery, Eye Glasses and Spectacles, Printing Presses, Flush Toilets, Microscopes, Telescopes, Submarines, and the Match. In turn, these inventions enticed the creation of economic value afterwards.

As noted in Chaps. 10 and 11, today's brain circulation is a very important driver of entrepreneurial growth. It refers to the flow of ideas, from clever young nationals who go abroad to study, then take a job abroad, and later bring back the fruits of that study and working experience to their home country. Some authors believe that this form of migration will increase in the future in some regions, especially if economic disparities between countries continue to diminish. Such a circular

Exhibit 91: Thinking Outside the Box: "Game Change": The Phoenician Longitude Navigation and Leadership

Phoenicians created a knowledge cluster in the Mediterranean basin. The value of the cluster was in its lateral dimension, which induced more brain in circulation rather than more brain drain.

Through their voyages, Phoenician traders, merchants, and thinkers motivated people to change. By longitude navigation (Edvinsson 2002),[2] the Phoenicians acquired a lateral perspective and thinking. This was a paradigm change that brought about a new mental model.

Lessons for today:

- International exposure through alliance partners
- Building bridges
- Cross-border projects between Industry and Academia
- Project-based international exchange of students and knowledge workers

migration, for example, has been observed amongst Indians and Malaysians who had studied in the United States and Australia, respectively. Notably, India, which has been deeply affected by the diaspora of brainpower that went in the direction of Silicon Valley, is experimenting with the new form of brain circulation that nourish tech entrepreneurship and fast moving start-ups in both nations.

Exhibit 92: The Indian Way to Brain Circulation

"India's younger generation moves to the United States for school, then takes their skills back to India, rather than building careers in areas like Silicon Valley – as Mayfield's Yogen Dalal and Telesoft Partners' Arjun Gupta did. Instead of expatriates, they are part of a wave of talented repatriates, Mr. Dalal says."

"Hundreds [of engineers] have returned to India since 2000 to start businesses or help expands R&D labs for the likes of Oracle, Cisco Systems, and Intel. The deeper, more symbiotic relationship developing between the [Silicon] Valley and India goes far beyond the 'body shopping' of the 1990s, when U.S. companies mainly wanted low-wage software code writers."

Source: Silicon Valley/San Jose Business Journal, December 12, 2003; Hof and Kriplani (2003)

Brain circulation is the most effective way of contrasting *brain drain*. In the 1990s, brain drain has been a vast phenomenon that has mainly affected the South of the world, with the United States having benefited heavily from this migration: "America's open immigration policies are one of the country's 'secret weapons' for supporting entrepreneurship. New immigrants start businesses at very high rates, and many of the US most successful businesses are headed by those born outside of the US" (National Dialogue on Entrepreneurship, February 9, 2004).

Exhibit 93: Brain Drain

It has been defined as "departure of educated or professional people from one country, economic sector, or field for another usually for better pay or living conditions." Brain drain weakens the weakest and reinforces the strongest.

Two main basic concepts regarding brain drain can be identified:

- Brain exchange
- Brain waste

Brain exchange implies a two-way flow of expertise between a sending country and a receiving country. When the net flow is heavily biased against the sending country, the term "brain drain" is used.

A further term, "brain waste," describes the waste of skills that occurs when highly skilled workers migrate into forms of employment not requiring the application of the skills and experience applied in the former job (Brzozowski 2007).

Brain drain is a part of a larger problem: the existing economical, technological, and social gap leads to a growing flow of immigrants from the South to the North. The number of people living out of their native countries rose from 75 millions in 1965 to 120 in 1990 and 150 in 2000. One third of the highly qualified personnel from Africa have migrated to European countries and that about 60,000 administrators, trained at the secondary school and university levels, have migrated between 1985 and 1990.

It has been estimated that only from Latin America 1.1 million scientists and researchers have migrated to industrialized countries during the last 40 years, meaning that there is a loss of 27,500 scientists and researchers per year. It has been calculated that the region has exported about US $30 billion during this period that corresponds to the cost of training of these scientists" (South Centre 2003).

"In the 1990s, roughly 650,000 people from emerging markets migrated to the United States on professional-employment visas. Over 40% of the foreign-born adults in the United States have at least some college education, thereby making that country the epicentre of the global talent drain. Foreign-born workers now make up 20% of all employees in the US. About 30% of the 1998 graduating class of the famed Indian Institute of Technology – and a staggering 80% of the graduates in computer science – headed for graduate schools or jobs in the United States. Around 80% of foreign doctoral students in science and engineering plan to stay there after graduation – up from 50% in 1985" (Devan and Tewari 2001). Students, mainly of Indian and Chinese origin, represent almost one third of the high-growth companies mapping America's entrepreneurial landscape.

Although the migration of high-powered intellectual assets enlarges the productivity gap, the international mobility of highly educated, talented young people helps to close that gap since it implies a two-way flow of entrepreneurship-led innovation between a sending country and a receiving country. This movement, in one way, holds off the phenomenon of brain waste. It also promotes the creation of "nations of entrepreneurs" dedicated to transferring ideas from university to market, and fostering collaboration between academia and the business community.

International student mobility provides exposure to communities of knowledge practice of the host institutions and facilitates the formation of boundary crossing multiple communities. Yet, student flows appear to be highly concentrated in a few countries and little eclectic in consequence of the historical, cultural, and linguistic ties, and the role of geographical proximity.

In the light of what we have pointed out above, there has yet to be developed an international framework that calls on the academic sector to play a more direct role in the shaping of a future international entrepreneurship policy, one dimension of which is international student mobility.

Universities and other higher education institutions together with companies and business associations should embrace the creation of an inter-cultural context of mobility and integration, opposed to a multicultural context of emigration and separation. Mobility rather than migration of high-powered intellectual assets stimulates the international transfer of ideas from the university to the marketplace and fosters international collaboration between academics and business people. A collaborative environment where participants can cultivate new business ideas and turn them into commercial realities would be fostering a spirit of entrepreneurship and innovation among today's youth while preparing them for the challenges of a global marketplace. Participants might move from one learning location to another, and, in each location, a big stimulus would be gained from the diversity and ethnic mix of both the student population and the faculty members.

The appropriate actions should be implemented through a combination of educational institutions and public and private sectors. This is a matter in regard to which responsibility has to be placed upon the shoulders of those who are responsible for changing the academic foundations on which human capital has been built during the machine age. The new foundations should set the stage for an innovative learning environment that epitomises the *clever polis* or the *knowledge city* of the twenty-first century renaissance. Here knowledge and skills are encouraged, the love of learning and an inquiring mind are fostered, creativity and imagination are emphasized, and a digital-connected collective intelligence is designed to maximize the creative collaboration of small groups of entrepreneurial people addressing issues that interest and concern them all.

The challenge is how to get a little organization formed by students who decide to pool their resources almost from the start to think like a global organization. By playing the role of matchmakers and thereby building networks of contacts with students, entrepreneurs, and venture capitalists around the world, education institutions operating in different countries may have much to contribute to the creation of small entrepreneurial student teams that are cross-cultural and cross-country.

Exhibit 94: International Student Mobility: OECD Findings

1. Student mobility is constantly growing.
2. Few OECD countries are taking off in terms of [student] internationalization. The phenomenon is very highly concentrated in few countries: only five countries, Australia, USA, France, Germany, and United Kingdom, attract 8 out of 10 foreign students of OECD area.
3. Taken overall, the analysis of the geographical distribution of students by countries of origin leads to the conclusion that the United States appear to have resolutely set its face toward the Asian continent and its neighbor, Canada, as has Australia vis-à-vis Asia. Students from China, India, Japan, and Korea make up a significant proportion of the total number of foreign students in the United States at nearly 40% of the total. In Australia, some 7 out of 10 foreign students come from Asia or the Pacific.
4. The directions of student migrations are governed by geographical, linguistic, cultural, historical, institutional, and academic considerations, (the latter, centers of excellence). Finally, student migrations are greater the higher the level of education, degree of technical content of the field, and the need for foreign students to master foreign languages (e.g., commercial disciplines).
5. Large countries, whose language is more widely used internationally, attract a larger number of students, especially those coming from one of the member countries of a regional economic group (European Union, NAFTA).
6. The very high mobility rate of European Union students (169 students per 1,000 students enrolled) in EU member countries is facilitated by policies on freedom of movement, recognition of degrees, and the existence of exchange programs and assistance for student mobility (e.g., specific programs such as Erasmus).

Source: Tremblay (2002)

Exhibit 95: Matchmaking by Education Institutions: A Vital Component for Multicultural and Cross-Border Start-Ups

"Business schools with highly diverse student bodies and international exchange programs, such as the International Graduate School of Management (IESE) in Barcelona and INSEAD, the European Institute of Business Administration in Fontainebleau outside Paris, are also spawning a new generation of multicultural start-ups. Tony Raad, a Lebanese-born British national, launched an e-commerce start-up called MedHermes this year with INSEAD classmate Thomas Lengweiler, a Swiss national. Both worked for 3Com in London before founding MedHermes in Paris, which offers medical professionals information, news, and interviews with experts. While everything on the company's first Web site is written in French, all the back-end technology is handled out of London and

(continued)

Exhibit 95 (continued)

operates in English. The pair knows how to work in multicultural settings: Raad's marketing team at 3Com included an Armenian, a Dane, a Greek, an Indian, an American, and several Brits" (Business Week November 20, 2000 INTERNATIONAL SPECIAL REPORT)

Students from the Norwegian School of Entrepreneurship take a course on Technopreneurship in National University of Singapore, offered by NUS Entrepreneurship Centre and work as interns at start-up companies in Singapore for a period of 3 months. They are expected to write a business case on the company they have worked with by the end of the program. They have been attending the graduate course TR5105 Technopreneurship in Norway via video-conferencing with the students in NUS. The TR5105 Technopreneurship course is offered as an elective to all Business, Engineering, Science and Computing graduate students at NUS. Students taking the course are formed into teams to create viable and scalable business ideas and develop them into business plans. Other student destinations are Boston University, Fudan University –Shanghai, and Cornell University – Silicon Valley (NUS-National University of Singapore e-newsletter, January-February 2004, Issue 9) (http://www.nus.edu.sg/nec/enewsletter/latest%20updates/Issue9/technoprog_04.htm)

International Entrepreneurship Across Ethnic Boundaries

Student mobility can serve the purpose of creating international start-ups that also show an intercultural dimension. This phenomenon could be accelerated by the current trends in the role models of entrepreneurship played by minorities.

In Europe, the University of Jönköping in Sweden has launched a pilot project for the integration of Islamic minorities through the creation of an intercultural network for start-up initiatives whose founders and team members would be students in international mobility, from different countries and cultural communities, enrolled in entrepreneurship education programs.

Tolerant attitudes toward minorities, immigrants, religion, and nationalism are the foundations for cross-border and across ethnic boundaries start-ups. According to the *Euro-Tolerance Index*, Scandinavia and Nordic countries "may have a distinctive competitive advantage in terms of tolerance." Contrast that with the situation in the United States, which "ranks second from the bottom, ahead of only Portugal and close to Ireland" (Florida and Tinagli 2004: 27). In the US, the number of minority entrepreneurs has grown sharply in recent years, but the more traditional and conservative attitudes of Americans with regard to several dimensions of tolerance appear at odds with a context conducive to the formation of international and intercultural start-ups.

Exhibit 96: Cross-Border and Across Ethnic Boundaries Start-Ups: An Experimental Laboratory for the Vitalization of Islamic Communities in Europe Based on Trans-National Cooperation

- A network of regions that proactively collaborate for the vitalization of their regional Islamic community by promoting collaboration between those European and Islamic universities and other institutions committed to the creation of international and intercultural start-ups.
- A network of regions that jointly share experience and develop methods and tools that are tangibly applied and evaluated in the local communities.
- Project based exchange of students and entrepreneurs – in particular, minority entrepreneurs.

Source: Henoch (2003)

Exhibit 97: Euro-Tolerance Index

The Euro-Tolerance Index combines three measures: Attitude Index, Values Index, and Self-Expression Index (see Florida and Tinagli 2004 – Appendix: Data and Methodology: 42–44).

1. Sweden 15.00
2. Denmark 12.09
3. Netherlands 11.42
4. Finland 9.49
5. Germany 9.45
6. Austria 7.76
7. United Kingdom 7.70
8. France 7.38
9. Belgium 7.35
10. Italy 7.17
11. Spain 6.57
12. Greece 5.58
13. Ireland 4.22
14. USA 3.07
15. Portugal 1.99

Governance Frameworks

In the new age of knowledge, information and access ("participation age"), emerging cross-border, transnational communities driven by innovation and entrepreneurship initiatives – in short, international entrepreneurial communities – give impetus to the rise of international public goods.

With varying intensity, a nonvoting international mobile public – still a small but an increasing fraction of the public as a whole – holds strong opinions on and seeks recognition of the concerted nature of international entrepreneurship as a new type of public good that requires international provision (Boettke and Coyne 2005).

The extent to which communities and populations are excluded from the benefits of this international public good will depend on the degree of social trust in the community as well as on its size and the quality of its entrepreneurially-oriented labor pools and complementary business activities. The higher the trust level, the lower the search costs and the higher the quality of available pools of complementary activities in a community, the more likely are its individual members to benefit from concerted entrepreneurship. While greater in size, in the form of a larger pool of expertise, it also has a favorable impact and it may at the same time carry the cost burden of increasing institutional inertia – thus emphasizing the advantages of optimal combinations of critical mass and flexibility.

Exhibit 98: International Public Goods

According to the International Task Force on Global Public Goods, "international public goods, global and regional, address issues that: (i) are deemed to be important to the international community, to both developed and developing countries; (ii) typically cannot, or will not, be adequately addressed by individual countries or entities acting alone, and, in such cases (iii) are best addressed collectively on a multilateral basis."

Source: http://topics.developmentgateway.org/poverty/rc/ ItemDetail.do|1069526?itemId = 1069526

There is a strong link between governance conditions and nonexcludability. Although with a different scale of intensity, excludability principles and practices permeated feudal "status" societies, later planned economies, and the mercantile "society of contract" of the manufacturing age. In many institutions, frameworks such as guilds or intradisciplinary codes of conduct created new barriers. The "society of relations" (the prevailing form in the participation age) is better suited to knowledge-driven entrepreneurship and, in particular, to its most advanced form, which relies on the concerting of international entrepreneurial developments.

Governance Modes 1 and 2

In a relations-based form of governance, which we call "mode 1," the primary role is given to behavioral rules that allow for communication and exchange to take place without rigid, formal contracts. Mode 1 leads to investment in social capital that ends or constrains restrictive practices at many levels, such as those that make the labor market inflexible or protect vested interests. Those restrictive practices lock individuals into particular roles, preventing them from adjusting in accordance with new business ideas brought in by concerted entrepreneurship activity.

Mode 1 allows space for new markets that promote transactions which generate positive externalities. Prediction markets – markets that aim to "support transactions in claims about unresolved questions of fact" (Bell 2006) – are a case in point. Prediction markets intervene to support exchanges that spread information about and develop consensus on unsettled questions regarding concerted entrepreneurship. Just as greater interdependence among different entrepreneurial players acts as a catalyst for a decision-making process that is conducive to the creation of international start-ups, so are cross-border business opportunities exploited more fully by a larger number of people in a relations-based community.

We call a rule-based governance regime "mode 2." Both modes 1 and 2 put in motion processes of "institutional competition" and of mutual influence between policy makers and emerging international entrepreneurial communities. The latter are subject to public regulations while "competing" with them by generating the social capital necessary to ensure that trustworthy relations grow. The combination of rule-based and relations-based governance is an indicator of the intensity of "institutional competition" between rule builders and relationship builders.

Cost Implications of Modes 1 and 2

A trustworthy community constrains the increase in marginal costs of relations-based governance – costs that inevitably occur with the enlargement of the concerted entrepreneurship zone (i.e., with increased numbers of participating entrepreneurs and relationships). The stronger the principle of nonexcludability in a community, the more likely it is that "distant" outsiders will participate. In dealing with each other for the purpose of new venture cocreation, insiders behave much as they would as with close neighbors, and the ability to establish strong trust-based relationships, which do not discriminate against outsiders relative to insiders, limits the increase in the marginal costs of enlarging the area of nonexcludability (Dixit 2006).

Mode 1 incurs high marginal costs in a low-trust community, which suffers from free-rider problems and the abuse of loose or noncontractual relationships. In this case, it is likely that a rule-based governance regime (mode 2) will prevail. Unlike mode 1, mode 2 bears high fixed costs because it requires a framework of legislation, regulation and formal contracts, and their enforcement by the courts. Once the rules have been fixed and are in place, however, negotiations with outsiders bear a low marginal cost (Dixit 2006). If a community cannot sustain extensive trust, excludability is the price it has to pay.

Institutional Steps

Communities differ widely in their levels of social capital and entrepreneurial endeavor. A serious focus on how modes 1 and 2 work, and under what circumstances they are effective, will send a powerful signal that policy makers are engaged in promoting competition between the two modes for the purpose of securing collective advantages from concerted entrepreneurship as an international public good (Fig. 22).

What exactly should a government do if it is set on, or forced into, a course of the mode 2 type? Since over-regulation stifles the development of open economic spaces and is thus an impediment to the full development of concerted entrepreneurship, policy makers must dispense with superfluous regulations and concentrate on striking a balance between the rules they enforce as "superior" decision makers and spontaneous actions by community members who, as contributors to the development of concerted entrepreneurship, must be taken into account in policy decisions.

To meet the conditions of nonexcludability, policy makers should focus on strategies and policies that favor the spread of concerted entrepreneurially-oriented efforts and opportunities. This calls for coherent action across a range of policy domains, spanning from early childhood training that plants the seed of entrepreneurial motivation and encourages the learning of foreign languages and openness to other cultures, to city planning, regional policies, the science–industry interface, and labor market institutions that help to create environments favorable to concerted entrepreneurship. If policy makers are committed and effective in their efforts to reduce exclusion, the vast majority of the community will reap the benefits of concerted in nature international entrepreneurship.

Policy Implications

International entrepreneurship spans cultural boundaries. Public policy has a role to play in developing an institutional environment that enables entrepreneurial action of the international entrepreneurship type. The basic and complementary cornerstones of a public policy favorable to the creation of start-ups that conduct business

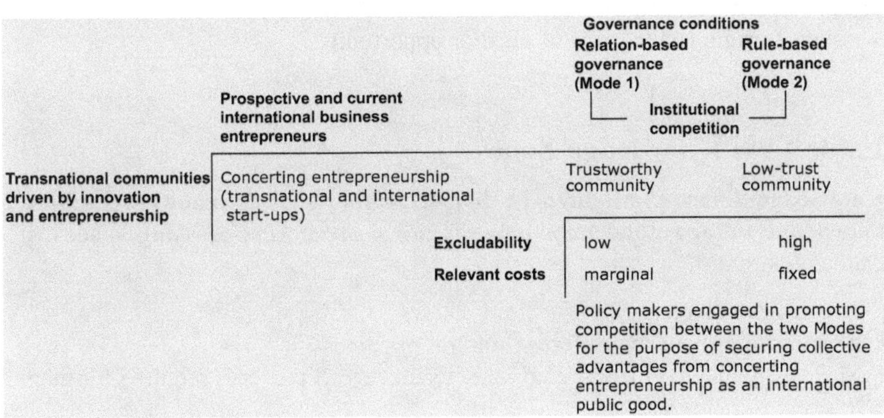

Fig. 22 Theoretical framework of institutional competition for international public goods: The case of concerted entrepreneurship

across borders are both the construction of a decentralized environment where people are free to move and a thoughtful concern for microeconomic policies that engender dynamic and boundary crossing communities. Because diversity is a key to effective cross-fertilization of ideas, in communities where cross-border ethnocentricity replaces local homogeneity, cross-country parties are inspired and energized by learning from one another about their distinct business environments.

Relation capital is the backbone of activities committed to cultivate an entrepreneurial culture that allows individuals from different geo-economic, cultural, ethnic, and linguistic environments to realize mutually beneficial opportunities. Inward and outward people mobility is a means of trying to ensure interpersonal relationships that lead to the collection of confident cross-border groupings with a shared mission, which is the creation of international start-up firms that can act upon opportunities for mutual gain that have been recognized during the mobility process.

Mobile students enrolled in entrepreneurship education programs portray one noticeable form of alertness to business opportunities for mutual gain. In this respect, the policy makers' mission should be that of organizing international student mobility by developing a common education space for project-based v exchange schemes, which result in student team internships with universities, higher education institutions, and companies from foreign countries. The provision of a common education space is a necessary but not a sufficient condition. That mission also portends budgetary consequences as fully portable grants and loans ought to be made available to mobile students to enforce student mobility agreements.

Working to open the door to the formation of knowledge zones is a second course of action taken by policy makers who are willing to respond to the needs of international entrepreneurship.

Knowledge zones are part of an emerging subset of microeconomic policy packages that policy makers should take into account and implement for the purpose of linking individuals who search for cross-border partners to convert into international start-ups ideas nurtured in their own communities. From this point of view, the main challenge of local policy makers is to build knowledge zones as the essence of their governance – which can facilitate the flow of business ideas from the point of origin to the point of need or opportunity.

Exhibit 99: Knowledge Zones

Knowledge zones, configured in diverse setting such as knowledge cities, knowledge villages, and knowledge regions, are border-crossing spaces of exchanges that:

- Are open to diversity
- Build bridges between different cultural practices
- Make available to citizens from other territories all the tools required for them to express themselves
- Foster connections, dialogue, and meaningful conversations

Notes

[1] Florida's measure of creative occupations includes professionals, artists, musicians, scientists, economists, architects, engineers, managers, and other workers whose jobs deal with creative, conceptual tasks (Florida 2004).

[2] Many eighteenth century ships lost their bearings since they could not navigate with any precision from east to west. The solution came from a knowledge outsider. In the years after the sextant was invented in 1731, many held out hope that it would aid in east-west navigation as well – that is, in finding longitude. Sailors could employ the sextant to figure out longitude using the lunar-distance method, but with the astronomical tables of the eighteenth century, the process could take several hours to work out one's position – not remotely good enough for sea travel. In the end, it was the dogged clockmaker, John Harrison, who solved the longitude problem with his chronometers. And today, the precocious stepchild of these highly accurate clocks is the Global Positioning System (Edvinsson 2002).

Chapter 14
Laboratory Experiments as a Tool in Empirical Economic Analysis of High-Expectation Entrepreneurship

There is no such thing as a failed experiment, only experiments with unexpected outcomes.
(Richard Buckminster)

It doesn't matter how beautiful your theory is, it doesn't matter how smart you are. If doesn't agree with experiment, it's wrong.
(Richard Feynman)

Experimentation for the entrepreneur will often focus on adoption of the innovation and the value that is created for both the end consumer of the innovation and the entrepreneur and the potential ecosystem that is required to deliver the innovation. For an innovation to be sustainable the innovation has to deliver value to the end consumer, the entrepreneur, and the innovation and delivery ecosystem, otherwise the innovation and entrepreneurial activity half-life will be short.

Experiments point out how high-expectation entrepreneurs should cultivate market outcomes, which behavior should guide trust building between the former and their potential financiers, and how policy makers should design and test "rules of the game." Persistent beta states (i.e., states associated with normal waking consciousness) for the business model and underpinning venture offerings become the norm. Rapid experiment iteration and rapid solution prototyping go hand-in-hand for the high-expectation entrepreneur, with plateaus of stability introduced to the iteration cycles, to enable commercialization and value capture from the evolving offerings.

The educational context under which high-expectation entrepreneurship could be cultivated would draw benefits from experiences made in the medical schools where different performance learning modes are created whose real impact is part of the educational and research activity. In particular, a business school should go beyond doing detached diagnoses, to really developing experiments, even of a therapeutic kind, and testing them clinically in interaction with private and public organizations.

T. Andersson et al., *Knowledge-Driven Entrepreneurship*, Innovation, Technology, and Knowledge Management 102, DOI 10.1007/978-1-4419-1188-9_14,

In Ireland, the Irish Management Institute has established a BizLabs program to support this kind of experimentation focused on broader organizational challenges. IMI BizLabs operates as a business think tank offering active communities of interest focused on specific organizational challenges working through

- Solution-driven research addressing current business challenges
- Industry-led collaboration with organizations to codesign solutions
- Ongoing research team investigation

Experiments, Simulations, and Clinical Treatments

Experiments

Defined as a "managerial behaviour which consistently exploits opportunities to deliver results beyond one's own capabilities" (Thompson 1999: 209), entrepreneurship requires enterprising individuals who can identify new opportunities and implement accordingly. Thus, entrepreneurship is a skill, learned through experience and improved with practice. With experience as the centerpiece for entrepreneurial development, the probability that entrepreneurs will learn from their experiences greatly increases.

Entrepreneurs continuously accumulate experience by conducting and evaluating experiments in the marketplace. Prior to their injection into the market process, inventing and innovating would-be entrepreneurs can find in economic experimenting labs a new *locus* for experimental activity. Experiences made inside the lab give rise to a range of views, which helps the decision maker limit his exposure to risk and uncertainty in the course of actions once field experiments must be carried out in the marketplace.

Laboratory experiments test propositions derived from new business ideas, which, on the basis of experience, need to be processed by monitoring and revising assumptions underlying and performance/reward predictions ingrained into those propositions.

In labs, participants learn the language of the market through the following:

- Conducting a test or investigation
- Direct observation of events (*events in process*)
- Participation in social interactions with their peers
- Placing the learner under realistic conditions in order to imitate or estimate how events might occur in a real company, in the related industry, in the marketplace, etc.

Learning from experience and implementing the experimental results are two essential steps high-expectation entrepreneurs have to consider for the purpose of reducing the level of risk intrinsic to new ventures focused on innovation.

Experimental results indicate policies to be developed, which can significantly reduce the start-up time. Less time needed for completing a start-up launch means that start-up costs are lower, much less the upfront capital that is required, and higher the probability that new start-ups are getting started. Moreover, experimental results trace the road conducive to interactions between established firms and experimental ventures. Through interactions, the latter could derive benefits from the accumulated experience of the incumbents in terms of accelerated innovation and growth.

Equally relevant is the reevaluation effect. On the basis of experience, the conclusion can be reached that experimental businesses have to change direction. Such an experiment-induced behavior, pursued by high-expectation entrepreneurs who see lab experiments as a potential stimulating way of evolving their business, is a source of advantage over the established firms, which interpret bad results as a consequence of underperforming managers rather than the outcome of wrong predictions.

Last but not least, since pattern recognition is an inherent feature of our nature as human beings, laboratory experiments must carefully consider the possibility to give ambitious entrepreneurial individuals seeking fast-growing business opportunities a chance to experiment and learn how to become pattern-completers. Those who underperform having difficulties along the process from pattern recognition to pattern completion can get help from lab clinics.

Exhibit 100: Events in Process

New product/service ideas are events in process, which call for a set of abilities in order to reduce the high level of risk intrinsic in the process from creativity to implementation of the business idea.
Laboratories performing experiments on business ideas put in motion a chain of events. Examples of events in process are as follows:

The business idea is absorbed as a nutrient into the body of the founding entrepreneurial team. Then, it is disseminated among the team members.

Knowledge prior to action is closely and dialectically interwoven with actions deemed pertinent to discover which facts are relevant to the business idea under investigation.

Consumers' preferences, which are unpredictable and subject to continuous transformations (von Hayek 1944), are submitted to a process of discovery – albeit imperfect it is for perfect rational decisions are often not feasible in practice (Simon 1957).

Simulations

Business simulators situate players in a virtual situation where they have to make decisions. Simulations push them not only to think, but also to understand how the real business world is, what they should keep in mind, and how their decisions affect the performance of a whole company.

The simulation process is an interactive learning method where the goal is to learn business by doing business in a risk-free environment. The learning player will have to make decisions to start-up and run his/her company according to different simulation stages: from the actual marketplace related to the business idea so that the simulation can run according to "reliable" data (like competitors, market share, sale of similar products, average pricing, alternative products, etc.) to the generation of random data by which to discover unknown happenings or scenarios in a way that is close to reality. It is important to note that simulators, as learning tools, increase performance, enhance the development of business and entrepreneurial skills, and reduce costs of making errors in real life. A correct interpretation of results can help to provide meaningful feedback to players so that they can adjust their efforts to master the subject matter. More than winning, the objective is to learn how to behave in certain circumstances; that is the real value of playing a "smart game."

Exhibit 101: Objectives of Simulations

To interpret correctly real-world (simulated) situations and train players' abilities to make and quantify decisions.
To be familiar with the creation of new ventures and all aspects involved
To carry out critical analysis of complex business interrelationships
To manage growth and organizational change
To gain clarity on the consequences of decisions to be taken
To draw up and systematically make use of objectives to run a company
To improve strategic thinking, link strategies to objectives, and then link these objectives to decisions
To improve teamwork and organization

Clinical Treatments

If experimental and simulative approaches try to prevent too dangerous and too expensive real errors to happen once start-ups are immersed in the actual workings of real-world markets, the clinical approach intends to heal the disorder encountered during experiments in labs or in living, taking on the role of therapeutic counsel who fights against its clients' diseases. Clinical treatments turn patients from passive and low expectation to active and high-expectation players, who challenge what business experts say, seek second opinion, and actively collect information.

Entrepreneurial Experimentation

By entrepreneurial experimentation we mean a method that relates a business concept to an experiment, which stimulates the concept creator to build upon that concept as it has been experimented in lab. Since the opportunity cost of experimentation decreases with the increase in the value of a new business formation, we can expect that high-expectation entrepreneurs whose high-value-added business propositions bear low opportunity costs reveal a propensity for conducting experiments higher than that of entrepreneurs pursuing low-value-added activities.

Would-be entrepreneurs start with beliefs and ideas they want to turn into a business. By running experiments, business ideas move from an embryonic stance to their full manifestation in the form of new ventures. In particular, conducting experiments gives potential entrepreneurs with high-expectation ambitions the opportunity to get the ability to mobilize their new ideas so as to anticipate changes, take a chance on the future, and organize the context that does let those ideas win access to the marketplace as reasonably successful business propositions.

Quantitative and qualitative data collected during experimental (and clinical) exercises make easier to ask for support from financiers, evaluate pros and cons of pure financial backers vs. strategic partners, seek industrial allies, and hire employees. Anyone familiar with high-expectation start-ups knows that the key to success is having access to the participation of experienced people who bring both talent and funding to the new venture.

Exposure Modes

Would-be entrepreneurs are agents exposed to modes of experimentation, which are either analogical or conjectural. The former accomplishes the task of shrinking the entrepreneurial agent's area of known ignorance about her/his business idea. Analogy-based reasoning can be applied to the known unknowns of a new concept whose business domain holds attributes that match with those of another domain. The analogical situation is represented by a typical case-based approach. Past cases from a different domain are used to choose possible solutions for problems incurred with the new idea.

The conjectural mode proceeds by trial (i.e., spontaneous, serendipitous discovery of building blocks of the business idea under experimental scrutiny) and error (i.e., elimination of arrangements that subsequently result inappropriate). High-expectation business concepts sail into uncharted waters as they exhibit unfamiliar traits of novelty and complexity. The entrepreneurial agent, who is unaware of her/his ignorance, is exposed to a voyage into the unknown unknowns. Thus, high-expectation propositions cannot be treated by analogical reasoning and, specifically, by case-based reasoning. When no apparent rules or commonalties can be applied, trial and error is the approach that can back the tasks required by an imagination- and conjecture-based process of discovery. The major cost of this approach is the time invested for yielding a solution from the iterative process triggered by selecting what ex ante looks like the most suitable choice set. If something does not work, the process has to be iterated until the appropriate answer is found.

Experiments in Collaboration: Agent-Based Experiments on the Nature and Perspective of International Start-Ups

Experiments in collaboration involve people keen to communicate across their knowledge boundaries. Main objects for collaboration are macroeconomic and social changes such as those brought about by the creation of international start-ups.

Until recently, global business was considered reserved for large multinational corporations and business conglomerates. Increasingly, however, there is a realization that entrepreneurial firms have a crucial role in international business (Mtigwe 2006). Indeed, the steady proliferation of internationalization has led to the inevitable development of entrepreneurship without borders (see Chap. 13). Experiments in collaboration concerning this particular atmosphere in which the entrepreneurial conditions for new venture creations are exposed to international tensions and pressure are a prologue to microexperiments focused on a given potential venture opportunity.

A kind of collaborative experiment on the nature and perspective of international start-ups has been carried out at the University of Tartu (Estonia) 2006–2007 master's class in entrepreneurship whose participants are would-be founders of new firms. Teams of 3–5 students, all engaged in content and context business processes aiming to new venture creations, have been at the heart of a collaborative experiment to test cultural values, psychological behavior, and business approaches to international start-ups. By running an experiment session, master's students earned introspection leading to search for conscious open mindsets along their processes of new venture creation.

The experiment made it clear that the ability to effectively manage and capitalize on international start-ups poses numerous challenges. Few start-ups whose founders are embedded in different countries and cultures may be successful, and even more may fail, but the knowledge required to flourish may not be found in textbooks, but rather in one's fellow classmates halfway across the globe.

Participants have pinpointed internal and external environmental conditions (Figs. 23 and 24) conducive to cross-border start-ups. Encompassing the facets of decision making, customer service, distribution, and communication, the internal environment of internationally founded firms proves promising for students. Citing

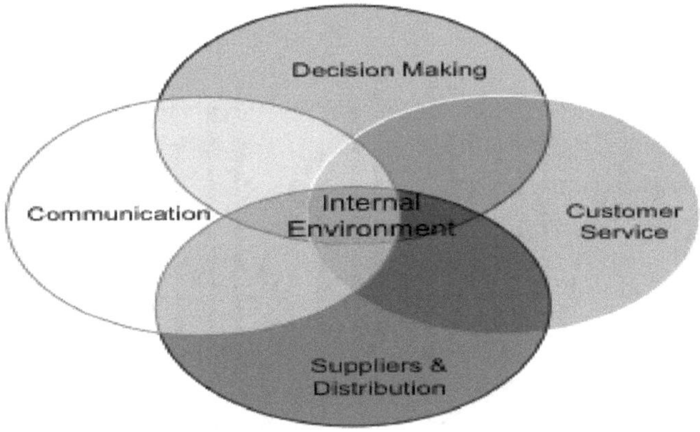

Fig. 23 Internal environmental conditions conducive to international start-ups

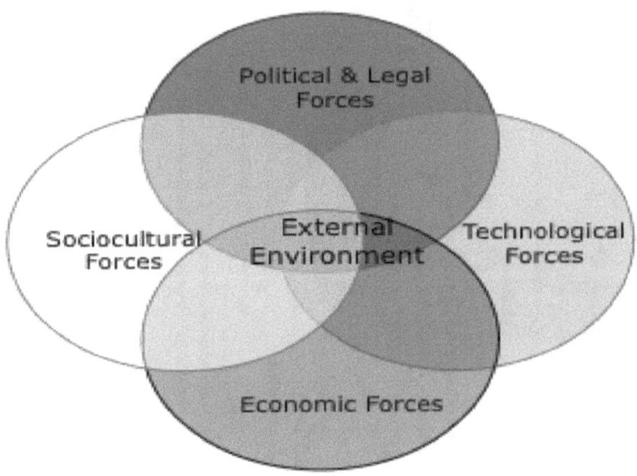

Fig. 24 External environmental conditions conducive to international start-ups

greater creativity and innovation, along with a better understanding of local values and customer demands leaves the benefits largely outweighing the shortcomings. Concern has emerged vis-à-vis delays in decision making and deliveries, along with miscommunication between employees.

As for the external environment, participants have easily identified the political and legal benefits of internationally founded firms. Other economic factors such as currency volatility have been less distinguishable. Concerns have been expressed regarding difficulties in integrating legal and political facets of each individual country along with cultural clashes between employees. These uneasy factors, however, have been largely outweighed by reassurance factors such as increased access to technology and access to foreign currency hedging.

Figure 25 highlights a mind map showing the overall results of the collaborative experiment conducted at the University of Tartu, in the pursuit of actions conducive

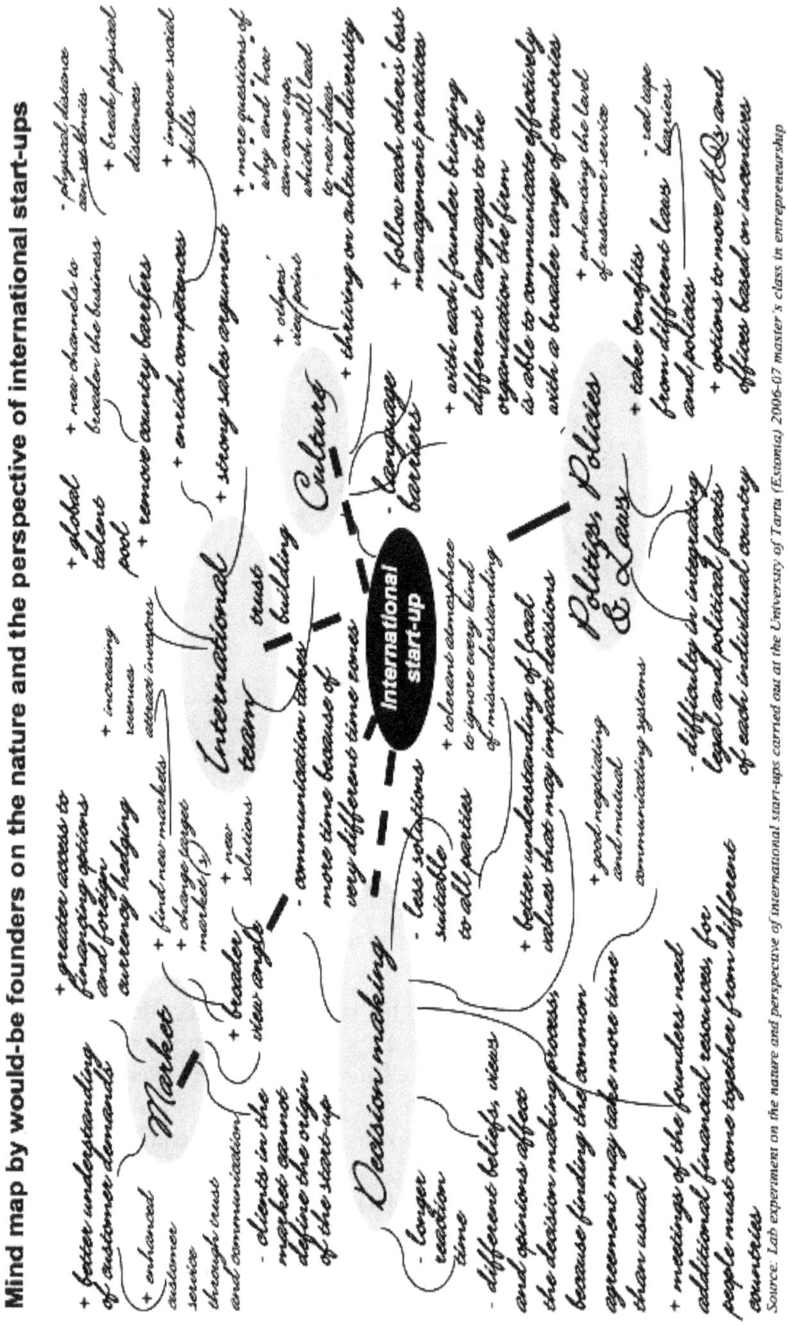

Fig. 25 Mind map by would-be founders on the nature and perspective of international start-ups

to new venture creation by means of international start-ups founded by master's students in entrepreneurship.

"Periodic Table" of Experimental Elements

An experimental lab, wherein business concepts are processed, can be looked through the kaleidoscope of its experimental elements. Each experiment consists of *n* experimental elements, which are organized in groups. We have identified five element groups: that is, Inputs, Process, Actions, Outputs, and Impact.

The five groups and their elements (we have identified 18 of them) are displayed in a "periodic table" of experimental elements. In Fig. 26, elements belonging to the same group are arranged in vertical lines. The proximity of a given element to another one depends on the intensity of their reciprocal interaction and influence. Higher the interaction and influence, closer they are. Since economic experiments compete with one another and each experiment is unique and unreplicable, for a given starting condition – that is, a given business concept – no deterministic law can be drawn as to the placement of the elements in the periodic table. Notwithstanding, experiment after experiment an adequate cumulative experience

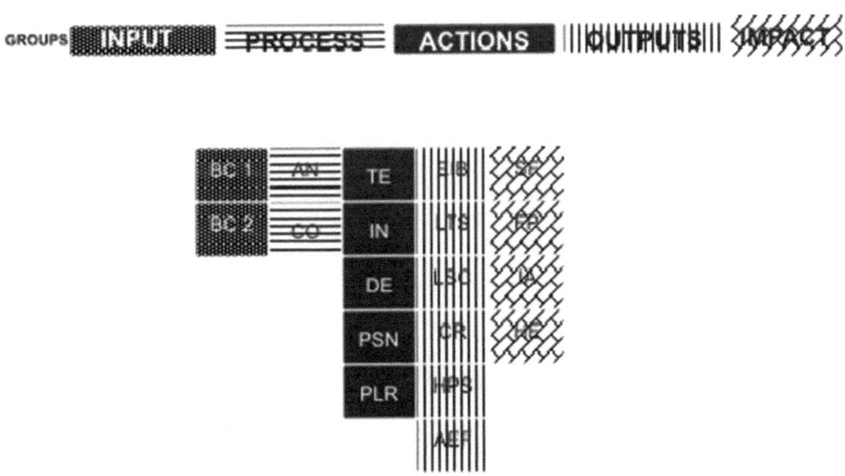

Fig. 26 "Periodic table" of experimental elements
BC: Business concept - AN: Analogical - CO: Conjectural - TE: Test - IN: Investigation - DE: Direct observation of events - PSN: Participation in social networks - PLR: Placing the learner under realistic conditions - EIB: Experiment-induced behaviour - LTS: Less time to complete a start-up launch - LSC: Low start-up costs - CR: Much less upfront capital required - HPS: Higher probability new start-ups are getting started - AEF: Source of advantage over established firms, which interpret bad results as a consequence of underperforming managers rather than the outcome of wrong predictions - SF: Support from financiers - FP: Evaluate financial backers vs strategic partners - IA: Seek industrial partners - HE: Hire employees

can be accrued, which results in probabilistic predictions of a given element's relative intensity in terms of interaction and influence.

The layout of Fig. 26 is one of *n* examples of how experimental elements interact and affect one another. Horizontal lines, called "periods," show two different experiments triggered by two different business concepts. Figure 25 can be refined and extended over time as new elements are identified, new groups are created, and new predictive models are developed to explain the entrepreneurial behavior over a given business concept process.

Appendix 1

Trends in the Industrial Age and the Knowledge Economy: A Tentative Glossary of the New Age of Knowledge, Information, and Access ("Participation Age")

When compared with the Industrial Age, the knowledge economy shows very different trends. Namely:

Industrial age	Knowledge economy
Valuing tangibles	Valuing intangibles
Trading physical products	Trading ideas and knowledge-based products and services
Optimization strategies: solving problems	Innovation strategies: seeking opportunities
Economic planning and forecasting	Prospecting future markets, responsiveness to needs and insights
Private ownership and controls	Shared ownership and co-opetition (combination of cooperation and competition)
Knowledge filtered by single experts	Knowledge filtered by communities of knowledge practice and their superior forms
Valuing technologies	Valuing people

Glossary of the New Age

Age of access: The age in which connectivity drives toward the access of everyone to everyone, everything to everything, and everything to everyone

Bionomics: The merger of biological and economic theory. In its more figurative sense, the merger of the world of the made and the world of the born

Building community: People investing in sharing content and sending messages to each other

Coevolution: Reciprocal evolutionary change in interacting species. Coevolution pushes competitors into "obligate cooperation." Alliances from cooperation are often asymmetrical for one party takes a greater advantage

Communication: The basis of culture, which is a process of communication among individuals and groups

Connectivity: The result of the fusion of computing and communication

(continued)

Content: A mere artifact of ability to communicate

Cyberspace: Communication as a destination in its own right, no more a pipe between physical locations on the planet. Our cyberspace identity is our email signature. Cyberspace is the mall of network culture. Cyberspace is naturally antisovereign

Infoeconomy: An environment in which atoms (products) and property of products have been replaced by bits (information) and sharing of information. Quicker the transmission of information, higher is the value

Knowledge landscape: An uneven landscape of empty know-nothing interrupted by hills of self-organized knowledge. Knowledge breeds knowledge as well as ignorance breeds ignorance. Knowledge processes consist also in mapping the holes of ignorance

Metering: Thanks to an information meter, everyone can buy what he likes to drink instead of an ocean of information. Therefore, metering converts information into a utility

Net: Less and less a thing and more and more an environment for higher resolution in each other communication. Net tends to grow organically – that is, not according to any person's conscious design, but because it is by nature a collection of individuals all making contributions to it

Netmarket: On the Net the marketplace is not divided into towns and regions, but into affinity groups. Noncommercial transactions are developed on the Net to foster a sense community

Netted intelligence: Networking of human intelligence through technology such as interactive multimedia and the so-called information highway

Network: A factory for information. As the value of a product is increased by the amount of knowledge invested in it, the networks that engender the knowledge increase in value

Source: Brockman (1997), Kelly (1994), Tapscott (1995), and Taylor and Wacker (1997)

Appendix 2

Emerging Technology Directions

As many technology companies emit more hype than Hollywood, it is no wonder that CEOs often struggle to identify a clear posture on technology. While Nicholas Carr's infamous article "IT doesn't matter" in the Harvard Business Review (Carr 2003) created a welcome debate, it is increasingly clear that technology is anything but irrelevant.

Technology is at the core of many of the changes in business which are happening. Harvard Professor Dale Jorgenson in 2001 emphasized the macroimpact of IT when he said, "Despite differences in methodology and data sources, a consensus is building that the remarkable behaviour of IT prices provides the key to the surge in economic growth" (Jorgenson 2001). Moore's law continues to power ahead with a doubling in capability every 18 months or so. This reignition of Moore's law, when conventional approaches to microprocessor design looked like dead-ending as chip temperatures began to approach rocket nozzle-like temperatures, has been driven by the adoption of multicore technology. A multicore microprocessor is one that combines two or more processors into a single physical package. Multicore technology not only improves performance but also improves so-called performance per watt. This is increasingly important as momentum for Green Computing initiatives increases. As we move out of an era of cheap energy and demand for computing and storage continues to grow, improving the energy efficiency of information technology will become of paramount importance. The movement of large data centers to locations with cheaper energy costs is the first data point in a new trend.

Among all the clutter of terms like SOA, Grid, Web 2.0, etc., a number of genuinely large themes are emerging which are of crucial importance to business and indeed society. The first is the convergence of computers and communications (every computer communicates and every communication device computes) enhanced by the emergence of ubiquitous mobile wireless communication. This theme can be likened to a "defining" technology such as the steam engine or electrification, which drove broad societal and business impacts and improvements. The evolution of ubiquitous mobile wireless communications enables new opportunities for both spatial and temporal mobility. Mobility in the temporal dimension means that individuals can allocate their time better, having more flexibility in completing tasks and also having the ability to consume more services. In terms of spatial flexibility, individuals no longer need to be tethered to their office to complete tasks or be involved in productive work. The concept of network centric enterprise is emerging which has implications for how individuals and organizations work to create value and change productivity, infrastructure, and culture relationships and meanings. The concept of network centric organizations has already been successfully applied in warfare leading to organizations which have better shared situational awareness and consequently information superiority.

In terms of achieving ubiquitous mobile wireless communication, probably the most exciting technology is Worldwide Interoperability for Microwave Access (WiMAX). WiMAX is a classic example of a so-called disruptive technology. The term disruptive technology was coined by Clayton Christensen (1997) to describe a new lower cost technology that displaces an existing sustaining technology – the IBM PC or the transistor replacing the Mainframe and valve technology, respectively, are other examples of disruptive technology. WiMax based on the IEEE 802.16 specification provides standards and technologies to provide carrier class communication solutions that can support hundreds of users with DSL-type connectivity as well as enabling T1-type connectivity to about 50 businesses using a single base station, at a cost much lower than conventional technology such as optical fiber or copper.

It is estimated today that there are more than 1 billion computers connected to the Internet but currently many of these connections do not have high-speed broadband connections. Broadband is a requirement for many of the new media-type services that are becoming available such as video on demand and online gaming. WiMax will likely provide the fundamental building block to connect the next billion users to the Internet and this will happen in a far quicker timeframe than happened for the first billion. A major standard and political obstacle to the adoption of WiMAX was overcome by the adoption of WiMax by ITU into the IMT 2000 set of standards. This enables WiMAX to coexist in spectrum typically reserved for 3G cellular systems and makes it much easier for governments and carriers to rollout widespread WiMAX networks. WiMax will operate as a complimentary solution to 3G and other cellular solutions rather than disrupt them.

As technology's relevance and pervasiveness increase, the role of the Chief Information Officer (CIO) will likely change. Indeed, in future CIO could stand for Chief Innovation Officer, or Career Is Over if the CIO fails at the new imperative around IT Innovation. There is a gravitational-like force driving the conventional enterprise Information Technology organization in the direction of utility like computing. Add to this the trend of the consumerization of IT where company employees can acquire much more powerful technology at their local consumer electronics store than the technology that is provided by their corporate IT department. Coupling this with the wave of growth around software as a service, virtualization, and grid computing means that many firms will in the future likely acquire their information technology plumbing from large hosting or utility firms.

Another emerging theme, which is fueled by information technology, is the emergence of services innovation. Services now account for about 70% of GDP and employment in OECD countries and many services are enabled or automated through Information Technology. Information Technology is often at the core of Services Provisioning.

In this context, IT or Services Innovation is emerging as a new discipline, one which exists at the intersection of two relatively immature disciplines, that of Information Technology and Innovation. While IT is increasingly being recognized as a discipline and the profession is reasonably well developed, it is just in the last couple of years that business schools have started to recognize that Innovation is not just something that happens by luck but perhaps is a process that can be mastered. The intersection of IT and Innovation as discipline creates the potential for accelerated value creation as IT is a unique innovation resource and it is also a useful resource for helping automate and manage the process of Innovation itself. Intel and other companies such as the Boston Consulting Group, Chevron, and Microsoft have established the Innovation Value Institute at the National University of Ireland to develop advanced frameworks and models to accelerate the maturation of IT innovation into a discipline in its own right so that organizations can more predictably capture and optimize the value from IT innovations.

The intersection of Mobility and Services Innovation indeed will lead to a new growth area which is location-based services. Location-based services are services which provide information specific to a particular location or sets of locations. This can include provide information to an individual such as the nearest restaurant or information to a firm to enable dynamic resource optimization by tracking and scheduling dynamic resources such as taxi or freight truck locations. When RFID technology is coupled to location-based services, further derivative service innovation is possible. Radio Frequency Identification (RFID) technology which uses radio waves to identify items is a fundamental enabler of what might be called an "Internet of things" where the world's objects are connected in both an intelligent and sensory fashion and an excellent logical representation of the physical world exists on the Internet. Integration of RFID-based solutions with breakthrough applications like Google Earth leveraging geographical information systems opens up opportunities for a whole new suite of potential services with privacy concerns the only limiter on what will be possible.

Web 2.0 is the name given to an emerging category of social networking tools, communities, and hosted services, which facilitate collaboration, sharing, and transactions between Internet users. The phenomenon of internal blogging within Corporations is bringing a new democracy to Corporations by enabling the free sharing of opinions, strategies, and thoughts in a more informal environment where employees can add or contribute to existing streams of conversation.

Virtualization is a critically important technology which represents the next frontier for infrastructure and applications. Virtualization can be implemented on a spectrum from micro- to macrovirtualization and is a way of logically presenting or grouping computing resources in a way which gives benefits over the normal physical configuration of resources. Microvirtualization occurs on a single physical machine such as a server or client, which can be partitioned into multiple environments called virtual machines (VM). Each VM can run its own operating system and applications. At the other end of the spectrum, all of the computing resources in an enterprise could be grouped as a virtual environment to run for example parallel applications or optimize overall resource usage. However, virtualization will not be limited to hardware and applications but will ultimately migrate to the individual enterprise and then the ecosystem.

Looking forward the accelerating development and emergence of new technologies presents both challenges and opportunities. CEOs should look to their CIOs and challenge them to play new roles in the enterprise, sourcing new emerging technologies and rapidly capitalizing on them to help create new products, services, processes, and business models. As the information intensity of many firms and ecosystems increases, how firms use information for competitive advantage will significantly determine longevity, growth, and sustainability and ultimately the world leaders in innovation will likely be the world leaders in everything else.

Appendix 3

Science and Technology Parks: Their Evolution

Adopting relevant imperatives from the economic and social environment is the type of evolution that science and technology parks (STP) need. We are already witnessing the evolution of the traditional science park models into new ones.

"Science Parks – argues David Rowe (2003) – started spent their early years creating infrastructure and buildings. This emphasis on property is hardly surprising. In the early years Science Parks had to establish what built environments worked well for the high tech sector and then huge energies were required to raise the substantial capital required for what was seen as a highly speculative and risky activity. But, by the early 1990s, it was time for the more adventurous to move on. The next step was to start taking more seriously the ways in which a Park could stimulate technology transfer between their associated University(ies) or centres of research and businesses on their Park, or perhaps develop ways assisting the development of start up and young high tech businesses in and around the Park or even stimulate the creation of new business support or research centres. The message from innovative Science Parks with strong business creation and support programmes became increasingly of interest. This gave Science Parks a significant role as a serious economic development actor for the first time and several leading Parks have thrown themselves strongly behind this role. The above evolutionary path is typical of many of the more successful Science Parks in Europe".

Luis Sanz (2003), Director General of the International Association of Science Parks, figures out what he calls the "Learning Village" as on the incoming technology-park model for the global society. This model comprises three main elements (1) businesses, (2) educational centers, and (3) residential areas, and their most important infrastructure will be IT based. The integration of these three elements will multiply exponentially the efficiency of Technology Parks. The presence of the latest IT-based infrastructure and its extensive, every-day-use, will enable its inhabitants to live, learn, and work in clearly identifiable geographical units, and yet be fully integrated into the new global society. Learning villages will be inhabited by so-called "globapolitans with roots." This model may also serve to ensure the interest and "'loyalty' of Science Parks' stakeholders, contribute to enhancing the role of Parks as social (and not merely economical) tools, and to ensuring its financial sustainability."

In a STP what is of most value is the combination of multiple stakeholders involved in it. These are competitors, partners, complementors, suppliers, and customers. Connectedness, that is bringing closer together providers and users of "innovation power," is a must. STP managers can develop connectedness through the formation of business communities centered round the consumer and communities of knowledge practice as well. Managers must also try to muscle in on the trend toward networking of human intelligence by expanding their role as a motor for the digital economy.

Companies would be ready to invest in STP for the purpose of acquiring insight and understanding of research once their capacity to assimilate advances in research was reinforced through entering into meaningful dialog with research institutions so as to fill the marketing and sales gap that hampers the transformation of inventions into marketable innovations. This is the very reason why the STP new generation replaces the linear model of transferring and its underlying law of unidirectional causality with the law of circular causality (Fig. 27). Nonlinear feedback loops link

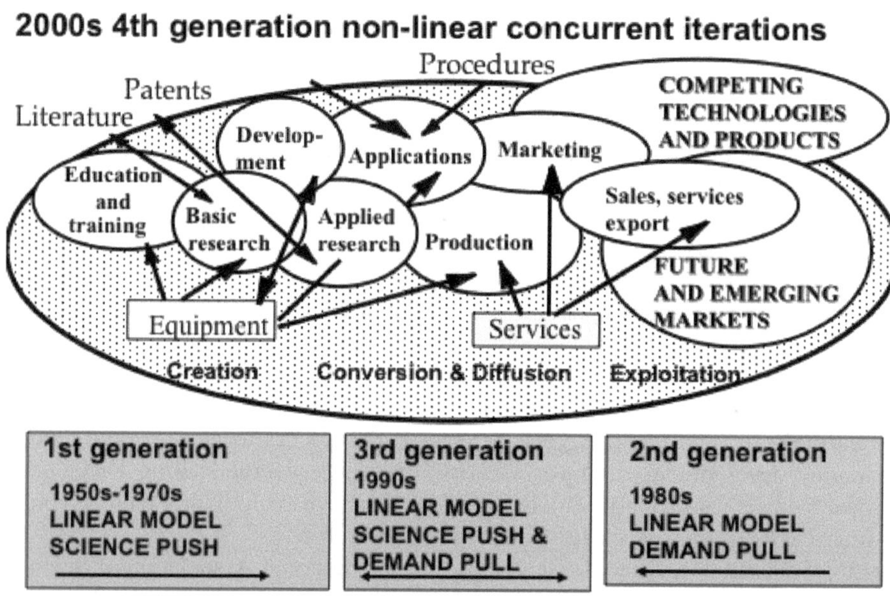

Fig. 27 Science park strategic context: the evolution The fourth generation of science parks moves away from the "1-mm wide, 1-mile deep" silos mindset of the science park functions, developed by the science park company in collaboration with its academic and business partners, tenants, and clients as well, toward a concurrent rather than linear process in which functions such as research, applications, production, marketing, and sales are simultaneously involved (Gann and Dodgson 2007).

The 4th generation of science parks moves away from the "1mm wide, 1 mile deep' silos mindset of the science park functions, developed by the science park company in collaboration with its academic and business partners, tenants and clients as well, towards a concurrent rather than linear process in which functions such as research, applications, production, marketing and sales are simultaneously involved (Gann and Dodgson, 2007) research to industrial innovation. A spiral model with a reverse flow from industry to research enhances the performance of the latter – which contributes in turn to amplify the virtuous cycle.

In the context of a market-driven transfer process, researchers, business strategists, and patent experts embedded in a STP coalesce in "invention-to-innovation teams," which are knowledge pools whose participants are accustomed to working together by following a Faraday-style behavior, for which "applied goals also tackle the basics."

Each team looks like a research enterprise whose "product" is a specific project with a limited lifetime (say, 5 years), unlike the bureaucracy of the conventional, age-old research institutes and laboratories where research projects often drag on for decades. At the end of the period, the project is discontinued and a new one with an entirely fresh team will take its place.

Exhibit 102: Science Park

The International Association of Science Parks (IASP) (http://www.iasp.ws) defines a science parks as "an organisation managed by specialised professionals whose main aim is to increase the wealth of its community by promoting the culture of innovation and competitiveness of its associated businesses and knowledge based institutions."

"To enable these goals to be met, a Science Park manages the flow of knowledge and technology among universities, R&D institutions, companies and markets; it facilitates the creation and growth of innovation-based companies through incubation and spin-off processes; and provides other value-added services together with high-quality space and facilities"

Appendix 4

List of Intangibles Based on the Financial Accounting Standards Board

(a) Marketing-related intangible assets

1. Trademarks, trade names
2. Service marks, collective marks, certification marks
3. Trade dress (unique color, shape, or package design)
4. Internet domain names
5. Noncompetition agreements

(b) Customer-related intangible assets

1. Customer lists
2. Order or production backlog
3. Customer contracts and related customer relationships
4. Noncontractual customer relationships

(c) Artistic-related intangible assets

1. Video and audiovisual material

(d) Contract-based intangible assets

1. Licensing, royalty, standstill agreements
2. Advertizing, construction, management, service, or supply contracts
3. Lease agreements
4. Construction permits
5. Franchise agreements
6. Employment contracts

(e) Technology-based intangible assets

1. Patented technology
2. Computer software and mask works

3. Unpatented technology
4. Databases
5. Trade secrets, such as secret formulas, processes, recipes

A *trade name* is any name other than the full first and last name(s) of the owner(s) of the business, including a general partnership.

A *trademark* is a word, phrase, symbol, or design, or a combination of words, phrases, symbols, or designs, that identifies and distinguishes the source of the goods of one party from those of others.

A *service mark* is the same as a trademark, except that it identifies and distinguishes the source of a service rather than a product.

Collective marks are usually defined as signs that distinguish the geographical origin, material, mode of manufacture, or other common characteristics of goods or services of different enterprises using the collective mark. The owner may be either an association of which those enterprises are members or any other entity, including a public institution or a cooperative.

Certification marks are usually given for compliance with defined standards, but are not confined to any membership. They may be used by anyone who can certify that the products involved meet certain established standards. Famous certification marks include wool mark which certifies that the goods on which it is used are made of 100% wool.

In many countries, the main difference between collective marks and certification marks is that the former may only be used by a specific group of enterprises, for example, members of an association, while certification marks may be used by anybody who complies with the standards defined by the owner of the certification mark. An important requirement for certification marks is that the entity which applies for registration is considered "competent to certify" the products concerned. Certification marks may be used together with the individual trademark of the producer of a given good. The label used as a certification mark will be evidence that the company's products meet the specific standards required for the use of the certification mark.

Noncompetition agreements are agreements between parties aiming for collaboration: for example, each party detains products and services that are complementary to those of the other party.

Appendix 5

Twenty Questions on Knowledge in the Organization: Results from a Survey by Ernst and Young

Twenty questions were addressed to executives within a broad industry coverage ranging from aerospace to utilities, from small businesses to large corporations. Most respondents (87%) described their businesses as "knowledge-intensive" and named multiple types of knowledge as being critical to their competitiveness.

Topping the list was "knowledge about customers," followed by "knowledge about best practices or effective processes, the company's own competencies and capabilities, and its products or services." Companies were strongly motivated in improving the intraorganizational transfer of existing knowledge and "facilitating knowledge through culture and incentives."

Asked what benefits from more active management of knowledge their organizations could gain, respondents often said "innovation." Forty-four percent rated themselves "good" or "excellent" in generating new knowledge leveraging on people.

Indeed, "an organization's knowledge advantage depends mostly on people" – respondents believe – as well as "upon people they put the emphasis about their organizations' ability to compete based on knowledge" (51% of the respondents).

As for the technology tools they believe offer the greatest potential for enhancing the knowledge base of their organizations, respondents said "the most initiatives undertaken to date involve Internet access and intranet, decision support tools, data warehouses, groupware, and directories of resident experts."

But the top three efforts were again people oriented: that is, "mapping sources of internal expertise; establishing new roles; and creating networks of knowledge workers." "Ultimately respondents-corroborated knowledge management comes down to people management."

Yet they "were fairly evenly split on the question of whether it would be valuable to create a new knowledge role, such as that of a 'chief knowledge officer,' to support people-related change." For some, the danger was too great that a staff position focused on knowledge would simply translate to more bureaucracy. "As sometimes happened with Quality – their argument goes – executives in charge might cease to focus ultimate business goals and pursue knowledge management for its own sake."

Respondents were aware of the major role of culture and people's behavior as the main vehicles for knowledge transfer in the organization.

Fifty-four percent rated "culture" as the number one impediment to knowledge transfer and 56% rated "changing people's behavior" as the biggest difficulty in managing knowledge within the organization. Also ranking in the top impediments to knowledge transfer was: second – top management's failure to signal its importance; third – the lack of shared understanding of strategy or business model, and the organizational structure; and fifth – lack of ownership of the problem. The report noted "that all of these are 'people' issues; technology limitations and non-standardised processes didn't make it onto the list. This is consistent with respondents' belief that knowledge management must be primarily concerned with people management."

Companies expected additional revenue from new ideas thanks to their ability to manage knowledge. Thirty-four percent of respondents to the survey mentioned "revenue generated by new ideas" as the "most useful measurement of knowledge performance" in their organizations.

Source: Ernst and Young Center for Business Innovation and Business Intelligence (1997).

Appendix 6

The Packaging Machinery Cluster in Bologna, Italy

Case studies of successful regions in Europe suggest that the viability of local economies relies on clusters of small- and medium-sized businesses that are a major platform ("wrestling school") for flexible, adaptable, and capable workforces. One of these success stories is represented by the packaging machinery cluster in and around Bologna, in the Italian region of Emilia-Romagna (Farrell and Lauridsen 2001; Formica 2001).

The Italian packaging machinery industry stands out internationally for its ability to meet the specialized needs of manufacturers throughout the world. At the same time, it remains the main supplier for Italian manufacturers. While about 85% of production is sold abroad, the industry is still able to fulfill more than 60% of demand at national level. Italian machines cover 25% of the world export market and Italian manufacturers are particularly strong in food, tobacco, and pharmaceutical machines, which accounts for over 70% of all equipment manufactured. The Italian trade balance in the sector is structurally positive in all major areas of production, including machines for cleaning, dying, labeling, filling, packing and packaging goods, as well as for the manufacturing of the individual machine parts.

Within the sector in Italy, we find companies large enough to offer a complete range of products to all world markets alongside smaller firms, which are able to fill in specific market niches. Both large and small companies provide the market with up-to-date, state-of-the-art technology. They are able to demonstrate a special sensitivity to the market needs of the manufacturers who use their services. Systems and machines are tailor-made to fit the specific needs of their customers, using innovative techniques, new packaging materials, or whatever else customers may require. That is why their agenda is full of crucial deadlines for innovation.

Exhibit 103: How an Industrial Pioneer and Two Probusiness Academics Built the Entrepreneurial Spirit in the Local Community

The packaging machinery industry in Bologna started with the foundation of ACMA-Anonima Costruzioni Macchine Automatiche on 31 July 1924, and has since then achieved a series of successes and major transformations in technology, research, and product quality (Fig. 28). In fact, almost all the companies in Bologna, dealing in packaging machines today, originated from that very first enterprise and from the workers and technicians trained by ACMA.

The founder was a partner in Gazzoni, a local pharmaceutical company that at the beginning of this century started the production of a powder to add sparkle to drinking water. Idrolitina: This was its trade mark. The powder had to be measured and packaged in paper by the hands of dexterous lady workers. In the early 1920s, as a result of a growing market for Idrolitina, the inventor, Signor Gazzoni, asked his associate, who ran a machine shop, to build a packing machine for the automatic packaging of table water powders. After this initial, successful experience, the ACMA machine production was enlarged to cover a wide range of chemical, pharmaceutical, confectionary, and food products. In the early 1930s, the company began to diversify into European and other foreign markets.

Several firms from 1937 have been spin-offs from ACMA. Their founders – artisans and technicians of the mother company – opened new market segments in producing automatic machinery and and a range of other products. As Farrell and Lauridsen (2001) have observed, "First, burgeoning cross-sectoral demand for packaging machines meant that there was a wide variety of market niches and room for many producers, and technicians could strike out on their own without succumbing to competition from their parent firm or other firms. Second, despite this variety, the mechanical skills needed to produce packaging machines for one market segment usually transferred with relative ease to another."

Today, in this industry in Bologna are a myriad of companies engaged in designing, making the machine parts (nuts, bolts, studnuts, washers, and many others), and assembly automatic machines for a wide range of industries, such as foodstuffs, bakery, confectionery, beverage, tea, tobacco, pharmaceutical, and chemical. Several of them are internationally renowned, and some are world leaders supported by chains of suppliers and subcontractors who deliver promptly in order to make the machine parts.

ACMA was the "food" molecule without which the "industrial reaction" would only have taken place with a great difficulty. The ACMA machines were conceptualised in the heart of the factory – the engineering department – directed by Bruto Carpigiani, the "father" of automatic machine designers from 1927 to 1945. He had acquired the mechanical knowledge disseminated through a local, commune-funded, technical school, Aldini-Valeriani, founded in the mid-1880s. Indeed, most of the spin-ees from the ACMA had been former students of the

(continued)

Exhibit 103 (continued)

Aldini-Valeriani. In the early days of the industrial revolution two Bolognesi – Giovanni Aldini, a scientist, and Luigi Valeriani, an economist, visited the new technical and professional schools in France, Great Britain, Germany, and Belgium, learning the best practice of the new technical education and training on offer in Europe. The fruit of their travel was first the gestation and then the foundation of a technical school. A self-sustained trend of new firm formation was the outcome of a cross-fertilization process between in-company learning by doing training and formal training at the technical school for new mechanical qualifications. Therefore, the Aldini-Valeriani School, in turn, has been the incubator of the catalytic process initiated by ACMA.

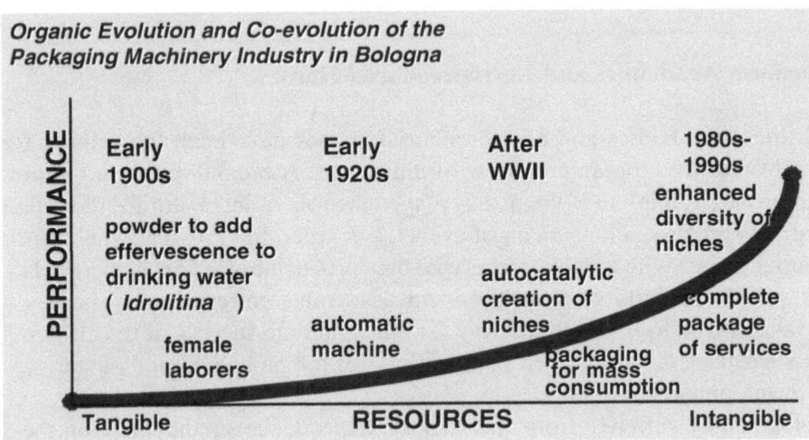

Fig. 28 From the first packaging machine to a complete package of services

Foundation of the Success: A Profound Technical Culture and a Widespread Entrepreneurial Spirit from Within the Business Community

Two trends that coalesce are responsible for the success of the Bologna packaging machinery industry. On the one hand, the great tradition in precision mechanics, which merged with entrepreneurial creativeness in a unique manner and paved the way for new industrial enterprises. On the other hand, the rapidly expanding demand for packaging in the market of large consumption products, and the need for adapting packaging machines to fit the particular requirements of each customer, so that even mass produced products include a "personalised touch." Both these elements have combined to assure quality production through careful specialization in all facets of manufacturing. The combination of the two trends gave birth to the packaging machinery cluster.

The cluster is an impressive case of collaboration between small- and medium-sized enterprises operating in continental or even global niche markets. The main industrial features that have contributed to them leading the field in this area encompass:

- Close vertical links between independent businesses, which have enhanced coordination as the prevalent type of interfirm collaboration
- Adoption of state-of-the-art technology
- Flexible production systems and methods
- Closeness to the customers

New trends include horizontal links between companies and a determination to work with partners for state-of-the-art research and development in a framework of cooperation and competition ("competitive cooperation" or "co-opetition").

Probusiness Academics and Entrepreneurial Heroes

Probusiness academics and entrepreneurial heroes have been the driving forces of a self-organized, organic process of clustering. A breakthrough in the field of education and training in Bologna has played the role of incubator for those heroes. Indeed, this process, a long chain of events, has lasted for 150 years, linking entrepreneurial heroes with another archetype: the "probusiness academic hero." Thanks to the social creativity of two prominent academics entrepreneurial heroes have materialized who have paved the way for the economic success of the cluster. This process has been driven by their personality over the pioneer period of start-up and early development (Exhibit 103).

Bologna has suffered from the lock-in dependence at the time of the first Industrial Revolution when the city was at the peak of its success in developing the wool and silk textile industry for which it was renowned in Europe as Lyon. Precepts and forms of the new scientific and technology domain, together with workforce practices that the Industrial Revolution introduced, caused a deep and prolonged recession of the Bolognese economy, influenced by a "Cargo Cult Science" of people, who had difficulty in understanding the principles of the incoming scientific age.

Nowadays, when the packaging machine industry is confronted with the leap from the industrial to the knowledge economy, from the production of "atoms" in form of machines that perform "cold" or unintelligent functions to that of "bits" associated with machines that even affect our very culture – in other words, from "making things" to "think-oriented, ideas-based businesses" – the reliance on the past success makes the cluster vulnerable to lock-in syndrome and "Cargo Cult." New heroes are heralded as essential to the mastering of a new domain. In the business-as-usual, good character actors replace the protagonists, but the shape of things to come is traced by the emergence of new leading personalities who are willing to change the fabric of traditional mental habits and conventional ideas supported by people with similar thought processes.

It is questionable whether new heroes can emerge from a successful industry. Well-regarded packaging companies have been attracting the best talents, but for new heroes this is not enough. The local community ought to demonstrate long-term commitment and willingness to create a new culture and invest in cooperative ventures. This carries far-reaching implications in the fields of research, education, and training. As already happened in Bologna, a fresh educational institution such as the entrepreneurial university could be the cradle of new heroes.

Appendix 7

Market Creation in the Entrepreneurial Economy: A Review of Online Markets

Great things can be accomplished by companies which create their own markets.

(Alan Cane)

Old-style industrial economies look like hi-end dinosaurs that mainly compete for existing, often conservative and mature markets. Producing more hardware than software, more machines and apparel than information, companies hang on the efficiency and effectiveness of their carriers to deliver atoms (Negroponte 1995). This has been the world of the industrial age. In the entrepreneurial economy, companies have to conceive ideas-based businesses and, therefore, they need to mold knowledge- and bit-intensive practices such as design, marketing, finance, and commercialization of innovation. Ideas-based businesses push companies to competitive cooperation ("co-opetition" – see Chap. 2). In this collaborative environment made up of competition and community, it happens that intense competitors become partners.

Permeated by the spirit of entrepreneurship, entrepreneurial economies are focused on markets not already in existence that could be missed. Increasingly modest gains arise at the horizon of conventional markets. The impending scenario of the Internet democracy for information and products (Exhibit 104) shows pictures of a deep change to the traditional approaches and thinking in the modern interdependent world. It is going to bring fresh competition in the free market economy and to narrow down the international borders.

Exhibit 104: Democratization of Information and Goods

- A continuous, rich, rapid, free, and reciprocal flow of information
- Reduced transaction costs
- A new marketplace, a new form of communication, and a new means of distribution
- Lower barriers to entry

Exhibit 104 (continued)

- Increasing price transparency and competition
- Easier price comparisons for buyers and sellers
- Cheaper prices of goods and services bought online
- Power shifted from producers to consumers
- A permanent increase in productivity and the level of output

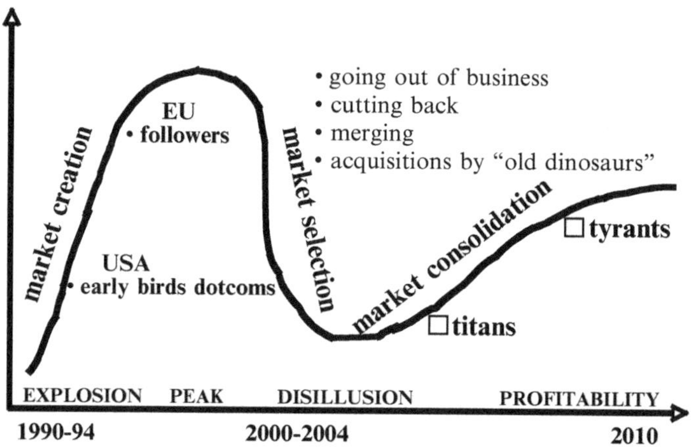

Fig. 29 Dotcom evolution

The spread of the Internet and the development of e-commerce might deliver significant cost reductions and organizational improvement to firms. Yet, the new weightless ways of creating wealth do not mean that novel markets lead to economic heaven. Indeed, for today's online ventures the road to the Promised Land of prosperity is navigation in troubled waters.

At the dawn of the Internet time early birds-dotcom companies have merged from the US to threaten values of the old economy's conventional wisdom. In the late 1990s, Andy Grove, the former Chairman of Intel, stated, "In five years time, all companies will be Internet companies or they won't be companies at all." This assertion has been deciphered as an imperative need and a challenge for all types of companies to use the Net "to lower costs, enter new markets, create new revenue streams, and, importantly, redefine relationships with customers and suppliers."

After have been plunging into the explosion phase till March 2000, a severe upheaval has struck the dotcom industry. A wider shake out echoes a trend toward the consolidation phase (Fig. 29). The aftermath of the wreckage left in the wake of the crash in Internet stocks has been that most of dotcom entrepreneurs who were not able or did not have time to raise enough money went away.

After the deluge, online markets have been reshuffled by the penetration of dinosaurs (how pure dotcommers dubbed old-fashioned bricks-and-mortar companies) who have been buying early birds dotcoms, as well as by the merging of

	Business	Consumer
Business	Business-to-business exchanges Companies purchasing on the web	Retail activity on the web: e.g. bookselling (Amazon), stockbroking (Schwab)
Consumer	Consumers bidding for prices and other features of goods and services: e.g. Priciline, Lastminute	Consumers' auctions: e.g. EBay

Fig. 30 The e-commerce matrix. Source: Shopping around the web. A survey of e-commerce. The Economist, 26 February 2000, p. 9

dotcoms survived to the wreck. However, the entire business population – incoming titans (that is, growth, "staging" firms that have been successful even in the creation of myths and symbols) and tyrants (that is, companies in the maturity phase) included – should not be comparable in size to the industrial giants of the recent past – see the paragraph about "Industry clusters in the web age," in Chap. 4).

Online Markets

Different species of Internet-based companies populate online markets. The most popular are those that sell a product or service to a retail customer (business-to-consumer or B2C dotcoms) and, then, those selling a product or service to another business (business-to-business or B2B dotcoms). *The Economist*'s E-commerce matrix displays four market segments: business-to-business, business-to-consumer, consumer-to-business, and consumer-to-consumer (Fig. 30). There are companies that offer solutions to end-users whereas others provide the foundations (the electronic bricks and clicks of Internet) on which the e-merchants can built their niches in the cyberspace. There are online businesses that rely on advertizing (for example, a portal site). Others become successful only to the extent that visitors make purchases.

Online markets can be built around buyers or around independent exchanges. The former markets are one-way networks in which big buyers hold the market power. The latter tend to be two-way networks that occupy a central position between buyers and sellers, mediating between both sides.

Companies already holding a great market power tend to reinforce their position by the attitude to enhance virtual markets through industry-led consortia that replace stand-alone online efforts. Inside an industry that has consolidated around a few large companies, these agree to use a Web site for the bulk of their B2B activities, thereby creating an industry consortium. This is also known as a "vertical portal" that is a "pyramid-shaped" biased market (Kaplan and Sawhney 2000), for it assembles a few big buyers used to work alongside a fragmented mass of small- and mid-size businesses which form different tiers of suppliers.

Exhibit 105: Industry-Led Consortia in the Pioneer Years of Online Markets

In 1999, General Motors, Ford, and DaimlerChrysler created Covisint, a consortium to reflect the fusion of Collaboration, Visibility, and Integration – one of the world's largest virtual market, which buy billion-worth of parts from tens of thousands of suppliers. Renault and Nissan Motor have agreed to join.

In 2000, big retailers Sears Roebuck of the US, France's Carrefour, and Sainsbury of the UK established a retail consortium, called GlobalNetXchange, which brought together $200 billion of annual purchases. Worldwide Retail Exchange, the GlobalNetXchange's main competitor, was founded by 11 members, including leading European retailers such as Tesco, Auchan, and Market & Spencer.

By contrast, online markets built around independent exchanges show fragmentation on both sides of demand and supply. They have been labeled "butterfly-shaped" neutral markets that are either focused on specific industries (named "pure vertical portals") or on specific functions and business processes across different industries ("pure functional portals"). Independent Net market makers or online intermediaries create this kind of portals within less consolidated industries where there are fewer big players commanding an important share of the market. With the balance of power split between several competing buyers and sellers, all participants share benefits from independent, neutral electronic trading.

The highest rewards ought to materialize in the most fragmented industries, those in which no more than 1–2% of the market is under the control of a single buyer or seller. The subjacent assumption is that liquidity, that is the volume of transactions with customer involvement, "depends on the basic level of fragmentation underlying each industry from the start. The more fragmented the industry, the greater the possibility of value creation for all parties involved" (Henig 2000).

Once a critical mass of buyers and sellers has been reached, vertical and functional portals allow buyers and sellers to find each other by means of a reduced number of searches and contacts (Fig. 31), thereby serving as electronic hubs.

	Without hub	With hub
Number of searches/ contacts to be undertaken	Each seller (S) has to search/contact each buyer (B). N = S x B if S = 5 and B = 5 then N= 5 x 5 = 25	The hub finds the buyers and sellers N = S + B if S = 5 and B = 5 then N = 5 + 5 = 10

Fig. 31 Matching potential buyers and sellers in a B2B market.
Source: Sawhney and Kaplan (1999)

Vertical and functional hubs are said to be complementors to each other. In fact, verticals lack functional expertise and functionals do not possess domain expertise. This is the reason why vertical and functional hubs tend to form "a patchwork of alliances" (Sawhney and Kaplan 1999).

There are also portals not involved in sales, acting only as providers of various kinds of industry/processes-specific information in the form of online bulletin boards, online journals, chat forums, etc., through which good and bad news travel quickly world-wide.

In today's online markets, information is a vehicle for buying and selling physical products. Trading of physical items means that online businesses are still under the shadow of a brick-and-mortar culture. They personify the Web's "hybrid" solutions.

A step forward is that of creating "pure" online markets thanks to "the ability of people to buy and sell pure information" – what has been coined "people-to-people" (P2P) online sector (Jacob 2000).

There are plenty of business opportunities for P2P online markets creation. Information transactions between people that generate revenue directly can be conceived for serious matters as the treatment of a disease or for trivial ones like the star-system gossip.

P2Ps do not bear heavy costs related to warehouses and delivering of physical products. Labor costs are also shaved since customers who generate contents are the bulk of their workforce. Yet, to really succeed P2Ps have to change customers' behavior. Nowadays customers are used to pay for information indirectly through the products bought online. P2Ps call for customers who ought to buy directly information and paying for what they have contributed to produce as content generators.

Buyers and Sellers: Who Is Who?

In the industrial economy, there is a clear divide between buyers and sellers. A swarm of unorganized individuals buy products manufactured by companies that hold a market power. Buyers tolerate inconveniences such as those caused by lack of information, uncustomized information, place of purchase, shipping delays, and, on top, prices fixed by sellers and cartels companies concoct to freeze price competition.

The landscape of digital economy shows a view on an alternative scenario. Boundaries between buyers and sellers are blurred. Acting as proactive, organized communities of like-minded people, buyers are no more restricted to the passive role of product and price takers. Thus, a web community of world travelers can determine the kind of flight and accommodation they are looking for, and offer a price. Sellers will bid to win that offer. A pool of customers can first collect information about them and then sell it for a given price to companies. A cluster of customers focused on a particular product or service can deliver personalized messages to a specific company as for the quality and the price of its offer, pushing the company to upgrade the current product/service (Davis and Meyer 1998).

Confronted to the challenge of customers who are no more passive recipients of what vendors decide to supply, companies are compelled to gain fresh insights to make better business decisions.

Search for Experience

A collection of digital technologies that shape the cyberspace might create an illusion of space. However, to the consumer's habit to searching for experience, e-companies cannot reply sending over the Net a mirage. Digital not less than physical companies need to look the customer in the eye, even though a digital one, and say: "I really want you to tell me what you think" (Feigenbaum 2000). Market creation in the cyberspace stems from companies that provide would-be customers with new experiences. The basic idea is that of building a Web site through which consumers of apparently similar goods can get an entire experience and learn their different features. Take the example of packaged goods such as tea, coffee, pasta, and look-alike. Their producers face two opposite pressures.

From one prospective, the big consumer goods companies make it very hard for a small company or a new one to beat their top-selling brands. Shoppers are used to buy the familiar brand, and so a large part of the price they pay is for the branding. The cost of advertizing and promotion required to come up with a competing brand would be much too high for a small company to cope with.

Another view is that, as a consequence of imitation and commoditization of those products at an accelerating pace, a growing number of individuals seem to be affected by the syndrome of "I-think-I'll-buy-that-instead," which means replacing the brand name with a product "just like it, but a bit cheaper." Indeed, the billion dollar brands' multinational companies are nowadays under the attack of indigenous brand owners, who have a better understanding of how to do business in their own countries, proliferating copycat products and supermarket's own-label goods.

The two conflicting strands could be managed, thanks to software tools, by e-companies capable of transferring an entire experience like that Starbucks successfully provides for drinking coffee. "Changing coffee drinking from a commodity industry to an emotional experience," Starbucks has created a new conceptual market. Starbucks sells "a retailing concept: the coffee bar, offering relaxation and conversation, and drinks made with quality beans, frothy and flavored milks, creams, syrups, and ices" (Kim and Mauborgne 1999).

Experience-led web companies are "personal shopping agents" who open up new markets by teaching consumers how to select, appreciate, and find a specific product or service. Personalization agents do not sell or ship any product or service. They are the consumers' trustworthy interfaces that organize a dialog with them based on an interactive e-mail technology through which contents could be added or modified, and new product ideas developed.

Summing up, changing the appeal of a product or service is what a personal shopping agent does really aim to succeed. To get success, online retail outlets need personal agents who will connect the consumer to the distributor. This is exactly what Amazon is doing. "Our core job – Amazon's Bezos says – is to help people make purchases, and we're investing more in personalisation that anybody else." Personalization and a strong brand are consistent with competitive prices (Exhibit 105).

A natural corollary of a process that helps people express their individuality by moving away from the mass market seems to be the fragmentation of the audience and, hence, of brands. It has been argued that global branding should loose its appeal in favor of local labels. There might be an even more shocking implication, which is that entry to a market would be facilitated because the user does not have to buy the brand at all.

With the needs of the individual dominating online markets, the recipe for success of e-ventures is based on "thinking about what makes people different, not what they have in common." But the search for experience by customized shoppers cannot be merely converted in a magic formula of adding a Web site to an existing business. A quantum leap is compulsory. The whole business has to be reshaped around the Internet's hi-charged features – namely:

- Communication easing
- Transaction costs approaching zero
- Each customer as a market segment of one
- One-to-one networking
- Price comparison easing
- Cost saving

Exhibit 106: Competitive Prices, Personalization, and Brand – The Amazon's Credo by Jeff Bezos

- Personalization and a strong brand are consistent with competitive prices.
- We are known for competitive prices. But we are also known for customer experience and great customer service.
- If your brand is based exclusively on price, you are in a fragile position, but if your brand is about great prices and great service and great selection, that is a much better position.
- There are a lot of different ways to construct a brand. I believe brands that stand for abstract concepts are more durable and more robust that brands that stand for something more concrete. A company's brand is a lot like a person's reputation.
- I would like our reputation to be that we start with the customer and work backward.

Source: Amazon.com and Beyond, Wired, July 2000.

Appendix 8

The Eight Dimension That Influence Knowledge Sharing

1. High power distance:
Power distance is the degree to which members of an organization (should) accept distinctions between members on the basis of organizational position.

2. High uncertainty avoidance:
Uncertainty avoidance is the degree to which members of an organization actively attempts to reduce ambiguity in organizational life by relying on norms, rules, and policies.

3. High humane orientation:
Humane orientation is the degree to which members of an organization encourage and reward individuals for being fair and kind in their interactions with other organization members.

4. High assertiveness:
High assertiveness is the degree to which members of an organization are assertive, dominant, and demanding in their interactions with other organization members.

5. High future orientation:
Future orientation is the degree to which an organization encourages and rewards long-term vs. short-term planning and projects.

6. High performance orientation:
Performance orientation is the degree to which an organization focuses on and rewards high performance and efforts to improve quality.

7. High individualism:
Individualism is the degree to which an organization focuses on individual accomplishment vs. group accomplishment.
Individualism is oftentimes a barrier that prevents information and knowledge exchanges. Individualism goes against team building and team performance. Under this circumstance, if the person who is pursuing an individualistic behavior leaves

the company then the organization loses his information and knowledge. If the influence of individualism is high, the qualitative evaluation is negative.

8. High organizational collectivism:

Organizational collectivism is the degree to which organizational members take pride in being associated with the organization.

Appendix 9

Three Groups of Intellectual Capital Indicators

The importance that a given company attaches to its people's capacity to act in various situations suggests that a first bunch of indicators should measure individual competence, which includes values, education, technical and social skills, and experience. Therefore, indicators must be constructed on the "land" of human capital. Namely:

- Reputation of key company employees with head hunters
- Years of experience within the profession
- Average number of years' service
- Recruiting costs of key employees
- Employee satisfaction, commitment, loyalty, enthusiasm, entrepreneurial spirit
- Proportion of employees generating new ideas and proportion implemented as new products and services
- Number of multifunctional project teams
- Value added per employee
- Levels of education and competence
- Number of people with PhD and/or master degree in percentage of total employees
- Expenditures on training and time in training
- Employee turnover
- Performance comparisons with competitors
- Linguistic and ethnic diversity

A second group of indicators is concerned with the measurement of the internal structure that consists of a wide range of patents, concepts, models, and computer and administrative systems. These are created by the employees and are thus generally "owned" by them. Namely:

- R&D expenses
- Number of patents
- Return on investment on the organization patents (perhaps taking into account cost of patent management)

- Income per R&D expense
- Project life cycle cost per $ of sales
- Average length of time for product design and development
- Target dates for regulatory submission (if applicable)
- Number of new product/service introductions
- Proportion of income from new product introductions
- Five-year trend of product life cycle
- Number of individual computer links to the database
- Number of times the database is consulted
- Contributions to the database
- Upgrades of the database
- Volume of Information System (IS) use and connections
- Cost of IS per $ sales
- Income per $ of IS expense
- Satisfaction with IS service

Around the external structure a third group of indicators is constructed. The external structure consists of relationships with customers and suppliers, brand names, trademarks and reputation, or image. Namely:

- Growth in business volume
- Proportion of sales by repeat customers
- Customer acquisition costs
- Customer satisfaction
- Customer complaints
- Brand loyalty and value
- Brands and trademark acquisitions
- Number of supplier/customer and/or R&D alliances and their value
- Revenue share from alliances
- Web-related revenues and inputs

Appendix 10

Two Cases of Creative Entrepreneurs: Enzo Ferrari and Lorenzo del Vecchio

Companies can be endowed with good engines. Yet, the creative drivers are those who make the real difference. Under the guidance of their creative founders, companies such as Sony, Ferrari, and Luxottica have been discovering new business paradigms and pursuing opportunities that break away from the familiar routines.

Morita's, Ferrari's, and del Vecchio's creative attitudes in business exemplify the *modus operandi* of the creativity map (Fig. 15).

Akio Morita turned technological islands (region 6) into new products for new markets and new clients (region 5).

Enzo Ferrari made to grow communities of practice that brought to the fore the importance of the regions 2 and 3 nearby the domain expertise

Lorenzo del Vecchio's has beaten competitors to market (region 4) by keeping a tight grasp over the distribution in accordance with a time-to-market strategy.

Enzo Ferrari: The Builder of Communities of Knowledge Practice

Over the 1930s, Scuderia Ferrari was a small, autonomous division of the Alpha Romeo Company. In 1947, Enzo Ferrari founded the company that bears his name in Maranello. In 2002, Ferrari was rewarded as the most respected Italian company in the world, according to a survey of more than 1,000 top managers in 20 countries across the globe.

In a personality driven context, Ferrari's legend extends well beyond the automotive world and motor racing and sports car industry, to reach the broader business community as well as the general public. The legend has been built around the efforts and determination of its founder and mentor, Enzo Ferrari. Although Ferrari was a good racer, his talent was in the direction of organization and the handling of small details.

Since the very early days of his education, Ferrari was driven by the tacit knowledge embedded in the field of his father's experience as a rural metal worker. His father's company, a small foundry, made sheds and gangways for the railroads in Italy. Ferrari was never interested in school. He had aspirations. One of these was to be a race car-driver.

Aged barely 20, Ferrari spent much of his time frequenting the Bar de Nord on Turin's Ports Nuova, getting to know people and making connections. He aimed at creating trust, fashion, roles, and maximizing the joint product of personal relationships within his small groups of peers, interacting informally out of the shop floor. Sharing and learning in the cafés, even playing cards rather than playing by business cards in the meetings: this was a common trait to the founders – in most cases, blue collars and technicians – of small companies in Italy.

Throughout the 1920s, Ferrari spent a lot of time judiciously creating his commercial and engineering connections. He also began surrounding himself with a group of close collaborators, including Gioacchino Colombo – the man who would eventually design the first Ferrari car after masterminding the Alpha 158s under Ferrari's patronage – and former Fiat technician Luigi Bazzi, a man who would survive into the 1960s as possibly Enzo's longest-standing lieutenant, having originally joined him in 1923. Bazzi had joined Alfa Romeo as long ago as 1922 after a spell in Fiat's experimental department, and would later become tagged as the man who conceived the fearsome twin-engined Alfa Romeo "Bimotore" in the 1930s.

Team Building by an Autocatalytic Process: Transforming Personal Knowledge into Organizational Knowledge

Ferrari "made" Bazzi and Bazzi "made" Ferrari. Not only Bazzi was a valued technical guiding hand, but also his long association with Enzo Ferrari enabled him to help smooth over the differences of opinion and temperamental problems, which made working with his boss an increasingly unpredictable, sometimes tempestuous, challenge in later years.

Through dialog and discussion, cognitive conflicts and disagreement were raised, which questioned the existing premises. This made possible the transformation of personal knowledge into organizational knowledge.

Long-standing, healthy rivalry has been at the origin of healthy collaboration – that is, collaboration to share complex information on an ongoing basis for a common goal.

During the time with Alpha Romeo, Bazzi was also responsible for tempting the highly respected engineer Vittorio Jano to leave Fiat to join the rival firm. Bazzi, who had also worked with Fiat, was at least partly responsible for persuading Ferrari that Jano was the right man for the job. Within months of joining Alfa, the ex-Fiat man was putting the finishing touches to the historic supercharged 2-liter P2, which made its competition debut in 1924.

Interpersonal collaboration across multiple boundaries – across cultures, functions, rivalries, and geography – featured in a mix of rivalry and cooperation between motor racing entrepreneurs, which gave birth, first, to knowledge pools and, then, to a knowledge cluster: the springboards for innovation through collaboration, rivalry, and creative imitation.

In a personality driven context, the key players were "strong heart" individualists endowed with a hedgehog-minded personality, who relate everything to a single central vision and focus maniacally on executing it. By raising rivalry but also building relationships among people, they made changes happened beyond the conventional wisdom horizon.

Bitter rivals to fellow Modena racing entrepreneur Enzo Ferrari were the Maserati brothers, the founders of Maserati in Bologna, Italy in 1914. Enzo Ferrari and Maserati brothers felt themselves reciprocally free when kept apart from one another in creating a new tier of intradomain business in the motor racing and sports-car industry.

Yet, the same forces that kept apart the founding fathers later on bound up the inheritors and successors. Once involved in relationships with one another, they were no longer free – they were part of the inexorable stream.

Today Ferrari owns Maserati and together they form a specialized industrial group that is unique in the world. The two companies compete in complementary market sectors with cars that have different characteristics. While Ferrari offers compact two-door coupe and spiders for street use, which find their origins of design in the advanced laboratory of Formula 1, Maserati works in a different way. Maseratis, with equal technology to Ferrari, offer performance that is less extreme and a level of comfort and everyday usefulness that allows it to stand out as an authentic grand touring vehicle of the highest level.

The Golden Handshake: "The Adam Smith's Invisible Hand of the Market Must Be Accompanied by an Invisible Hand Shake"

Over the past five decades, Ferrari and Pininfarina have had the world's best-known and most influential association between an automotive manufacturer and a design house. Battista "Pinin" Farina has been the creator of the Italian style in the architecture of the automobile. In the 1930s, he founded "Carrozzeria Pinin Farina." His plan was to build special car bodies.

Though Enzo Ferrari and Battista Pininfarina yearned to work with each other in the early 1950s, the road to real collaboration was hesitant to start. "Ferrari was a man of very strong character," Sergio Pininfarina recalls. "Therefore, Mr. Ferrari was not coming to Farina in Turin, and my father was not going to visit him in Modena, which was approximately 120–130 miles away. So they met halfway in Tortona."

That fateful rendezvous would alter the world's automotive playing field. "Everything became extremely easy once they sat down at the table" – Pininfarina continues. "They never spoke about any type of price. Both were very enthusiastic, for each thought, 'This will be great,' it was. 'I will give you one chassis, and you will make one car.' The first steps were tentative, much like two outstanding dancers being paired for the first time. The initial effort yielded a handsome perfectly proportioned 212 Inter cabriolet that had its official public debut at 1952's Paris Auto Show."

Lorenzo del Vecchio: The Proactive Market Maker

Leonardo Del Vecchio is the founder of Luxottica Group, the world leader in the design, manufacture, and distribution of prescription frames and sunglasses in the mid- and premium-priced categories.

The company was established in 1961 at the foot of the Dolomites, in Agordo, around the industrial district for spectacles. Yet, the company's imprinting is quite the opposite of those firms embedded in industry clusters. The Luxottica's economic performance has not been depended upon the creative imitation emphasized by the way of working within the industrial districts in Italy, but it has relied on innovative competition, particularly of the creatively destructive type, emphasized by Schumpeter.

The propensity to run a race in which a big payoff for his innovative effort is secured to the winner has been cultivated inside the knowledge pool inspired by Del Vecchio. This pool recalls the entrepreneurial spirit of the old-style workshop rather than the bureaucracy of the industrial company.

Having been educated at the Brera Academy of Art in Milano to study drawing and engraving, Del Vecchio's profile is that of the Renaissance man, a blend of artist and artisan, or, in the metaphor of a contemporary business leader, Nobuyuki Idei, chairman of Sony, that of an "orchestra conductor who is hearing all the instruments, but is listening for the overall sound."

So, the founder and his peers in the knowledge pool share an entrepreneurial mindset that prevails on the business-as-usual attitude.

This explains why Luxottica has taken decisions that have been defeating the existing premises in its industry, such as:

- Vertical integration instead of outsourcing. Luxottica manages all significant components of the eyewear production and distribution process. Vertical integration has allowed the company to produce quality eyeglasses with the lowest production costs and the highest margins in the industry.
- Direct instead of indirect distribution. Luxottica Group is not only a first-class manufacturer but also a proactive market maker. Its distribution network is operative in major markets worldwide through 29 wholly owned wholesale subsidiaries. A strong presence in the retail business has been achieved through successive, forward-looking acquisitions: in 1995 the acquisition of LensCrafters, the largest optical retail chain in North America; in 1999 that of Bausch and Lomb's sunglass business, which includes brands like Ray-Ban® and Revo®; in 2001, Sunglass Hut International, a leading sunglass retailer with over 1,900 stores worldwide.

- Stock market quotation to raise accountability and image. The paternalism, which is a widespread sentiment in the family-owned companies, prevents them to become publicly traded companies. The few who decide to be listed in the stock market do this to attract risk capital. Luxottica has been an early pioneer among the family companies of the industrial districts in Italy in going public internationally (listed on the New York Stock Exchange on January 23, 1990), and not because cash problems but as a matter of accountability and image.

Notes

1 Moore's Law as originally stated was that the doubling of transitory density every couple of years, which translates to higher performance for roughly the same manufacturing costs.

2 According to research from TheInfoPro, an industry analyst, the average installed capacity in Fortune 1000 companies has jumped from 198TB in early 2005 to 680TB in October 2006. This equates to a doubling in capacity every 18 months.

3 Ferrari History: http://www.thecollection.com/new/maserati/history.htm
History of Ferrari in Formula One: http://home.clara.net/nigelk/history.htm
Interview with Sergio Pininfarina, Automobilia, Milano, 1997: http://www.pininfarina.it/eng/history/cooperation/ferrari2.html
Crow JT (1981) Ferrari's early years. Road & Track 44
Levin DP (1988) Enzo Ferrari, Builder of Racing Cars, is Dead at 90. New York Times, August 16: Section D23
Nre, Doug, An Appreciation of Enzo Ferrari. In: Prova on-line Ferrari magazine. http://www.prova.com/ (look under "Editorials")

4 "The Greek poet Archilochus says: 'The fox knows many things, but the hedgehog knows one big thing'. The hedgehog's vision is of one, of a single substance. The hedgehog is a monist" (Berlin 1953).

References

Accascina G (2005) Information and communication technologies for development in the Arab States. United Nations Development Programme (UNDP), New York

Acs ZJ, Varga A (2004) Entrepreneurship, agglomeration and technological change. Paper presented at the 1st GEMS research conference: entrepreneurship, government policies and economic growth, Berlin, 1–2 April, 2004

Amazon.com and Beyond: Catching up with Jeff Bezos. A conversation with Wired Editors, Wired, July 2000, pages 252–257

Amidon DM (1997) The emerging community of knowledge practice. In: Knowledge, Inc.'s executive report on knowledge, technology and performance, March issue

Amidon D (2001) The innovation superhighway. Entovation International News, No. 49

Amidon D (2002) The innovation superhighway. Butterworth-Heinemann, Boston, MA

Amidon D, Bryan D (2004) Knowledge innovation zone (KIZ) – Principles. http://www.inthekzone.com/principles.htm

Amidon D, Formica P, Mercier-Laurent E (eds) (2005) Knowledge economics: emerging principles, practices and policies. Tartu University Press, Estonia

Andersson FCA (2000) Corporatism and economic performance. Working papers, Department of Economics, Lund University, No. 21

Andersson T (2008) The future of universities in Europe. In: The future of the research university, meeting the challenges of the 21st century. Kauffman Foundation, Kansas City, MO

Andersson T, Johansson de Silva S (2006) Ethiopia: innovation and growth in international comparison. IKED, Malmö

Andersson T, Hansson E, Schwaag-Serger S, Sörvik J (2004) The cluster policies whitebook. IKED, Malmö

Andersson T, Jacobsson A, Mossberg A, Sörvik J (2006) Enabling trust in the digital world, exploring the market for authentication technologies in international digital transactions. IKED and GTC, Malmö

Arthur WB (1989) Competing technologies, increasing returns and lock-in by historical events. Econ J 99(394):116–131

Aubert J-E, Reiffers J-L (2003) Knowledge economies in the Middle East and North Africa: toward new development strategies. World Bank, Washington, DC

Babson College (2003) GEM report findings. http://www3.babson.edu/Newsroom/Releases/IntlGEM2003.cfm

Bailey R (2000) The law of increasing returns. The Cato Institute, Washington, DC, March 18

Baldwin E, Curley M (2007) Managing IT innovation for business value. Intel Press, Santa Clara, CA

Baumol WJ (2002) The free-market innovation machine. Princeton University Press, Princeton, NJ

Baumol WJ (2004) Education for innovation: entrepreneurial breakthroughs vs. corporate incremental improvements. In: NBER working paper, April 30

Beck U (2003) Toward a new critical theory with a cosmopolitan intent. Constellations 10:453–468

Beinhocker ED (2006) The origin of wealth. Evolution, complexity, and the radical remaking of economics. Harvard Business School Press, Boston, MA

Bell T (2006) Prediction markets for promoting the progress of science and the useful arts. George Mason Law Rev 14:37. http://ssrn.com/abstract = 925989

Berlin I (1953) The hedgehog and the fox. Simon & Schuster, New York

Bhidé A (2000) Looking back to the next century. In: Closing keynote address, TIE annual conference, May, mimeo

Boettke PJ, Coyne CJ (2005) Concerting entrepreneurship: an international public good. In: Boyd G, Rugman AM, Padoan PC (eds) European-American trade and financial alliances. Edward Elgar, Cheltenham, pp 199–226

Brandt R (2003) The global village program, mimeo

Brockman J (1997) Digerati. Encounters with the cyber elite. Orion Business, London

Brzozowski J (2007) Brain waste, educational investments and growth in transitional countries. In: Working paper. Cracow University of Economics, Poland

Buchanan JM (1965) An economic theory of clubs. Economica 32:1–14

Bughin J, Choi V, Johnson, B (2008) The next step in open innovation. McKinsey Q, June

Carayannis EG GWU School of Business Lectures on Science and Technology, 1996–2005

Carayannis EG, Alexander J (1999) 2Winning by co-opeting in strategic government–university–industry (GUI) partnerships: the power of complex, dynamic knowledge networks". J Technol Transfer 24:197–210

Carayannis EG, Allbritton M (1997) A case study of computer-mediated communication among 100 scholars in 15 countries. Online Journal of Internet Banking and Commerce. http://www.arraydev.com/ commerce/JIBC/9806-07.htm

Carayannis EG, Campbell DFJ (2006) Knowledge creation, diffusion and use in innovation networks and knowledge clusters: a comparative systems approach across the United States, Europe and Asia. Praeger, Westport, CT

Carayannis EG, Formica P (2006) Intellectual venture capitalists: an emerging breed of knowledge entrepreneurs. Ind Higher Educ 20(3):151–156

Carayannis EG, Piero Formica P (eds) (2008) Knowledge matters: technology, innovation and entrepreneurship in innovation networks and knowledge clusters. A comparative, socio-technical systems approach in the US, Europe and Asia. Palgrave MacMillan, Basingstoke

Carayannis EG, Sipp C (2005) e-Development towards the knowledge economy. Palgrave Macmillan, Basingstoke

Carayannis EG, von Zedtwitz M (2005) Architecting GloCal (Global–Local), real–virtual incubator networks (G-RVINs) as catalysts and accelerators of entrepreneurship in transitioning and developing economies. Technovation 25:95–110

Carayannis EG, Ziemnowicz C (eds) (2006) Re-discovering Schumpeter. Palgrave Macmillan, Basingstoke

Carr N (2003). "IT doesn't matter." Harvard Business Review. MA: Harvard Business School Press, May

Casas-Pardo J, Puchades-Navarro MA, Sajardo-Moreno A (2001) Volunteer nonprofits' provision of collective goods as a public choice decision. In: IUDEM working paper

Chatman JA, Polzer JT, Barsade SG, Neale MA (1998) Being different yet felling similar: the influence of demographic composition and organisational culture on work processes and outcomes. Adm Sci Q 43:749–781

Chesbrough H (2003a) Open innovation: the new imperative for creating and profiting from technology. Harvard Business School Press, Boston, MA

Chesbrough H (2003b) How to generate more innovation for less. Financial Times, August 7, Thursday

Chesbrough H et al (2008) Policies for open innovation: theory, framework and cases. In: Research project funded by VISION Era-Net, Helsinki, Finland, July 2008

Christensen C (1997) Innovator's dilemma. Harvard Business School Press, Boston, MA

Coase RH (1990) Firm, the market and the law. University of Chicago Press, Chicago, IL

Coleman JS (1990) Foundations of social theory. Belknap Press of Harvard University Press, Cambridge, MA
Cox LW (1997) International entrepreneurship: a literature review. College of Business Administration, Florida International University, mimeo
Curley M (2002a) Peer-to-peer computing enabled collaboration. In: Proceedings of international conference of computation science, Amsterdam, LNCS 2330. Springer, Berlin
Curley M (2002b) Peer-to-peer, Okonomische, technologische und juristiche Perspektiven. In: Schoder D et al (eds) Peer-to-peer. Springer, Berlin
Curley M (2004a) Peer-to-peer pressured IT. Knowledge Management Magazine–ARK, London
Curley M (2004b) Managing IT for business value. Intel Press, Oregon
Curley M (2007) The IT capability maturity framework. Innovation Value Institute, Maynooth
Davis S, Meyer C (1998) BLUR – The speed of change in the connected economy. Capstone, Oxford
de Ferranti D, Perry G, Lederman D, Maloney W (2002) World Bank Latin American and Caribbean studies viewpoints from natural resources to the knowledge economy. World Bank, Washington, DC
Devan, J, Tewari, P S (2001), Brains Abroad, The McKinsey Quarterly, No 4
Dingwall R (1996) Professions and social order in a global society, University of Nottingham. Paper prepared for plenary presentation at ISA Woking Group 02 Conference, Nottingham 11–13 September
Dixit A (2006) Evaluating recipes for development success, World BANK policy research working paper 3859. World Bank, Washington, DC
Djeflat A (2002) Knowledge economy for the Mena region national systems of innovation in the Mena region. World Bank, Washington, DC
Downes L, Mui C (1998) Unleashing the killer app – digital strategies for market dominance. Harvard Business School Press, Boston, MA
Drucker P (1954) The practice of management. Harper & Row, New York
Drucker P (1969) The age of discontinuity. Harper & Row, New York
Drucker P (1993a) Innovation and entrepreneurship. Harper Collins, New York
Drucker PF (1993b) Management: Tasks, responsibilities, practices. Harper Business, New York
Drucker P (1993c) Post-capitalist society. Harper Business, New York
Drucker P (1997) Post-capitalist, Wired 1.3. http://www.wired.com
Drucker P (1999) Management challenges for the 21st century. Harper Business, New York
Economic Commission for Africa (2002) Economic report on Africa 2002 – tracking performance and progress. Addis Ababa, Ethiopia
Edmondson G (2000) International report. Business Week, 20 November
Edvinsson L (2002) Corporate longitude. Navigating the knowledge economy. Financial Times/ Prentice Hall, Upper Saddle River, NJ
Ernst and Young Center for Business Innovation and Business Intelligence (1997) Twenty questions on knowledge in the organisation – survey
European Commission (2004) Helping to create an entrepreneurial culture: A guide on good practices in promoting entrepreneurial attitude and skills through education. Directorate-General for Enterprise, Brussels
European Commission (2008) Future internet portal. http://www.future-internet.eu/. Last accessed Jan 2009
Eustace C (ed) (2003) THE PRISM REPORT 2003. Research findings and policy recommendations. European Commission Information Society Technologies Programme, Report Series No.2, October, Brussels
Farrell H, Lauridsen A L (2001) Collective Goods in the Local Economy. The Packaging Machinery Cluster in Bologna, mimeo (also in: Crouch C, LeGalès P, Trigilia C., Voelzkow H. (eds.), Local Production Systems in Europe, Oxfor: Oxford University Press, Vol. II
Feigenbaum AV (2000) Remember the customer. Financial Times, March 6, Monday, p. 11
Feynman RP (1999) The pleasure of finding things out. The best short works of Feynman, R. P. Penguin Books, London

Flake GW (1999) The computational beauty of nature: computer explorations of fractals, chaos, complex systems, and adaptation. MIT, Cambridge, MA

Florida R, Tinagli I (2004) Europe in the creative age. Carnegie Mellon Software Industry Center, co-published in Europe with DEMOS, February. http://www.demos.co.uk/catalogue/creativeeurope_page370.aspx

Formica P (2001) Frontiers of Business Collaboration. The Case of the Packaging Machinery Cluster in Bologna Region, mimeo (a previous version with the title "A Voyage to Innoland" has been published in: Formica P, Taylor, D (eds.), Delivering Innovation: Key Lessons from the World-Wide Network of Science and Technology Parks, Digital Malaga: IASP Publication, 1998)

Formica P (2003a) Industry and knowledge clusters: principles, practices, policy. Tartu University Press, Estonia

Formica P (2003b) Enzo Ferrari: the making of the motor racing and sports car knowledge cluster. Case study of an inspired leader. Men's Higher College of Technology, Abu Dhabi, mimeo

Formica P (2003c) Corporate governance of cluster development agencies: The case for market orientation. In: Bröcker J, Dhose D, Soltwedel R (eds) Innovation clusters and interregional competition. Springer, Berlin

Formica P (2005) The argument for international entrepreneurship in the knowledge economy. In: Structural change in Europe 4, entrepreneurial spirit in cities and regions. Hagbarth Publications, Germany

Fukuyama F (1995) Trust: the social virtues and the creation of prosperity. Free Press, New York

Gadiesh O, Gilbert J (1998) Profit pools: a fresh look at strategy. Harvard Bus Rev 76(3):139–147

Gann D, Dodgson M (2007) Innovation technology – How new technologies are changing the way we innovate, NESTA, Provocation 05, September, London

Garavelli AC, Gorgoglione M, Albino V Strategies for knowledge transfer: transmission and acquisition. Politecnico di Bari, Italy, mimeo

Garavelli A C (NOT AC!!!!) Gorgoglione M, Albino V (n.d.), Strategies for Knowledge Transfer: Transmission and Acquisition, Mimeo, Politecnico di Bari

Gilder G (2000) Telecosm. Free Press/Simon & Schuster, New York

Gilder G (2002) Moore's quantum leap, Wired, January

Gissing B (2005) Improving science–industry–regional authority collaboration. In: Formica P, Stabulnieks J (eds) Knowledge based entrepreneurship. EffeElle Editori, Cento-Ferrara

Gladwell M (1999) Clicks and mortar – Don't believe the Internet hype: the real e-commerce revolution happened off-line. Annals of Retail, December 6

Gladwell M (2000) The tipping point. How little things can make a big difference. Little, Brown and Company, Boston

Hague D (1991) Beyond universities: a new republic of intellect (Hobart Papers). Institute of Economic Affairs, London

Hall PA, Soskice D (eds) (2001) Varieties of capitalism: the institutional foundations of corporate advantage. Oxford University Press, Oxford

Halperin A (2001) Globally dispersed startups – Executive summary of research conducted at the Sloan School of Management at MIT

Hampden-Turner C, Trompenaars F (1997) Mastering the infinite game. How East Asian values are transforming business practices. Capstone, Oxford

Handy C (1995) Gods of management: the changing world of organisations. Arrow Books, United Kingdom

Hay C (2002) Tough questions for Europe. Financial Times, August 2, Friday

Hayek FA (1944) The Road to Serfdom (ARK edition 1986). Routledge and Kegan Paul, London

Hellman J, Jones G, Kaufmann D, Schanderman M (2000) Measuring Governance, corruption, and state capture: how firms and bureaucrats shape the business environment in transition economies. World Bank, Washington, DC

Henig PD (2000) Revenge of the bricks. Red Herring Magazine, 1 August

Henoch B (2003) Intercultural proactive network for start-up initiatives and vitalisation of Islamic communities in EU. School of Engineering, Jönköping University, mimeo
HMSO (2003) Lambert review of business–university collaboration: final report, London
Hof RD, Kriplany M (2003) India and Silicon Valley: now the R&D flows both ways. Business Week, December 8
Horibe F (2002) KM and innovation: can they thrive together? Knowl Manage Rev 5(2):18–21
Howard WG, Guile BR (1992) Profiting from innovation. Free Press, New York
Hunt J (1998) The value of conflict. Financial Times, December 16, Wednesday
Hutton W (2007) Britain should flaunt its knowledge. Financial Times, November 12
Idei N (2000) The law of increasing returns. Interview released to Dan Gillmor, political leaders also need to change, future company, October 13
Imaginatik (2004) Deep dive: a world of ideas – the benefits of diversity in idea management. Corporate Innovation, April, http://www.imaginatik.com
ISTAG (2009) Revising Europe's ICT strategy, EU information systems and technologies advisory group
Jackson T (1998) Melding of minds to master the intangibles. Financial Times, June 15, Monday
Jacob K (2000) Entrepreneur in residence. Red Harring, October
Jarillo JC (1993) Strategic networks: creating the borderless organisation. Butterworth-Heinemann, Oxford
Jeffry T (1999) America's entrepreneurial revolution: the demise of brontosaurus capitalism. Babson College, Wellesley
Johnson JE (2004) Factors influencing the early internationalisation of high technology start-ups: US and UK evidence. J Int Entrepreneurship 2(1–2):139–154
Jorgenson D (2001), "Information Technology and the U.S. economy". American Economic Review
Joseph K (2002) Growth of ICT and ICT for development realities of the myths of the Indian experience, discussion paper 2002/78. UNU/WIDER, Helsinki
Kaplan S, Sawhney M (2000) B2B E-commerce hubs: towards taxonomy of business models. Harvard Bus Rev 78:3
Karam JT (2001) Review of "Flexible Citizenship: The Cultural Logics of Transnationality" by Aiwha Ong (Duke University Press, 1999). Anthropol Q 74(1):45–46
Katz J (2002) Investment, finance and enterprise development programme for South America. OECD Steering Group Meeting, Buenos Aires
Kauffman S (1995) At home in the universe: the search for laws of self-organisation and complexity. Oxford University Press, Oxford
Kay J (1995) Foundations of corporate success. Oxford University Press, Oxford
Kelly K (1994) Out of control: the new biology of machines. Addison Wesley, Reading, MA
Kelly K (1998) New rules for the new economy. Fourth Estate, London
Kilpi E (2005) Why we simply need to get better at working together: Theoretical considerations behind management challenges today, mimeo, Helsinki
Kim WC, Mauborgne R (1999) Emotional blend of coffee. Financial Times, May 29, Thursday, p. 26
Kirzner IM (1985) Discovery and the capitalist process. University of Chicago Press, Chicago, IL
Klein K (1996) The economics of ideas (interview to Paul Romer), Wired, June
Kluver R, Fu W (2004) The cultural globalisation index. http://www.foreignpolicy.com/story/cms.php?story_id=2494
Kornai J (1992) The affinity between ownership and co-ordination mechanisms: The common experience of reform in socialist countries. In: Poznansik Z (ed) Constructing capitalism: the re-emergence of civil society and liberal economy in the post-communist World. Westview Press, Boulder, pp 97–116
Kosko B (1994) Fuzzy thinking. Harper Collins, New York
Kurzweil R (2001) The law of accelerating returns. http://lifeboat.com/ex/law of accelerating returns. Last Accessed Jan 2009

Leadbeater C (2000) Living on thin air: the new economy. Penguin Books, London
Leonard DA (1998) Mining knowledge assets for innovation. Knowl Manage 1(1), August–September
Liebowitz S (2002) Re-thinking the network economy, AMACOM
Linton P, Schuchhard P (2009). The digital world in 2025, European Internet Foundation Policy Report.
Lloyd-Jones S (1994) Corporatism and the development of corporatist ideologies in Spain and Portugal. cphrc Portugal's history online, http://www.cphrc.org.uk/essay/article9.htm
Locke R (2002) What is American corporatism? FrontPageMagazine.com, September 13
Martin BR (2003) The changing social contract for science and the evolution of the university. In: Geuna A, Salter AJ, Steinmueller WE (eds) Science and innovation: rethinking the rationales for funding and governance. Edward Elgar, Cheltenham
Martin R, Sunley P (2002) Deconstructing clusters: chaotic concept or policy panacea? Paper presented at the high technology small firms one-day clusters conference, Manchester Business School, 18 April
McDougall PP, Oviatt BM (2003) Some fundamental issues in international entrepreneurship. Kelly School of Business, Indiana University, July, mimeo
McGrath RG, MCMillan I (1995) Discovery driven planning. Harvard Bus Rev 73(4):44–54
McInerney F, White S (2000) Future wealth. Investing in the second great wave of technology stocks, Truman Talley Boohs. St. Martin's Press, New York
Means G, Schneider D (2000) MetaCapitalism: The e-business revolution and the design of 21st century companies and markets. Wiley, New York
Miles RE, Snow CC (1992) Causes of failure in network organisations. California Manage Rev 34(4):53–72
Milton-Smith J. (2002), In Search of Silicon Asia: Strategies for Developing Technopreneurs in the Asia-Pacific Region. In: Formica, P., Sanz, L. (eds.), Frontiers of Entrepreneurship and Innovation: Readings in Science Park Policies and Practices, IASP, Malaga
Mitra J, Formica P (eds) (1997) Innovation and economic development: university–enterprise partnerships in action. Oak Tree Press, Dublin
Moore JF (1993) Predators and prey. A new ecology of competition. Harvard Bus Rev 71(3):75–86
Moore GA (1999) Crossing the chasm: marketing and selling high-tech products to mainstream customers (revised edition). Harper Business Book, New York
Morita A (1992) 'S' does not equal 'T' and 'T' does not equal 'I'. The first United Kingdom innovation lecture. Royal Society, London
Moritani M (1982) Japanese technology. Simul Press, Tokyo
Mtigwe B (2006) Theoretical milestones in international business: the journey to international entrepreneurship theory. J Int Entrepreneurship 4(1):5–26
Nairn G (2000) Winds of change sweep through the marketplace. Connectis July(Issue 4):11–17
Nalebuff BJ, Branderburger AM (1996) Co-opetition. Harper Collins, London
NCOE – U.S. National Commission on Entrepreneurship (2002), Entrepreneurial Culture: Talking About Culture and Entrepreneurship, No. 52, Tuesday, June 11 http://www.redherring.com/insider/2002/0409/2509.html
Negroponte N (1995) Being digital. Alfred A Knopf, Inc., New York
Nicolò V (1996) Technology transfer through connection with stimulating markets. In: Formica P, Guedes M (eds) The economics of science parks. ANPROTEC, IASP and AURRP, Rio de Janeiro
Nonaka I, Takeuchi H (1995) The knowledge-creating company: how Japanese companies create the dynamics of innovation. Oxford University Press, Oxford
OECD (2001) Special edition of science, technology and industry outlook: drivers of growth – information technology, innovation and entrepreneurship. OECD, Paris
OECD (2002) Conference on the measurement of social capital, Rapporteur's summary, London, 25–27 September
OECD (2003) ICT and economic growth. Evidence from OECD countries, industries and firms. OECD Science, Technology and Industry Directorate, Paris
OECD (2005) Economic survey of China, Paris

OECD-DATAR (2001) Local networks of enterprises in the world economy: sixteen forums, Paris, January 23–24
OECD/Government of Canada (2003) The opportunity and challenge of diversity: a role for social capital? Synthesis report, international conference, Montreal, Québec, 23–25 November
Ottaway M (2001) Corporatism goes global: international organisations, NGO networks and transnational business. Global governance – Carnegie endowment for international peace, September
Paterson I (2003) The god of the machine. Transaction Publishers, New Brunswick, NJ (originally published in 1943 by GP Putman)
Perroux F (1950) Economic space: theory and applications. Q J Econ 64:89–104
Perroux F (1955) Note sur la Notion de Pôle de Croissance. Economie Appliquée 7:307–320
Pescovitz D (1996) The future of schools. Wired, June
Piatkowski M (2002) The 'new economy' and economic growth in transition economies, the relevance of institutional infrastructure. Discussion paper 2002/62, UNU/WIDER
Pigato MA (2001) Information and communication technology, poverty, and development in sub-Saharan Africa and South Asia. Africa region working paper series. World Bank, Washington, DC
Porter ME (1998a) On Competition, Harvard Business School Press
Porter ME (1998b) Clusters and the New Economics of Competition, Harvard Business Review, November-December, Vol.76, No.6
Porter ME (2000) Location, Competition and Economic Development: Local Clusters in the Global Economy, Economic Development Quarterly, 14, 1
Powell JA (1987) Is design a trivial pursuit? Design Stud 8(4)
Powell J, Harloe M, Mike Goldsmith M (2000) Achieving cultural change: embedding academic enterprise. Paper presented to the IMHE conference, 'Beyond the Entrepreneurial University? Global Challenges and Institutional Responses'. OECD, Paris, 11–13 September
Prince C, Beaver G (2001) The rise and rise of the corporate university: the emerging corporate learning agenda. Int J Manage Educ 1(2):17–26
Prince C, Beaver G (2002) Redefining the role of the corporate university: a UK perspective. Industry & Higher Education, August, Vol. 16, No 4
Rand A (1943) The fountainhead. Bobbs-Merrill, Indianapolis, IN
Reid I (2002) What is needed to make Australia a knowledge-driven and learning-driven society? Position paper no. 5 prepared for the Business/Higher Education Round Table, Melbourne, June
Robinson J (1960) Imperfect competition revisited – the causes of monopoly. In: Collected economic papers, vol. II. Basil Blackwell, Oxford
Rowe D (2003) Evolution applies to science parks too. In: Formica P, Sanz L (eds) Frontiers of entrepreneurship and innovation. Readings in science park policies and practice. IASP, Malaga
Sala-i-Martin X, Artadi EV (2002) Economic growth and investment in the Arab world. Discussion paper 020308, Columbia University, Department of Economics, New York
Sanz L (2003) Learning village: from technology parks to learning villages: a technology park model for the global society. In: Formica P, Sanz L (eds) Frontiers of entrepreneurship and innovation. Readings in science park policies and practice. IASP, Malaga
Saul JR (1995) The unconscious civilisation. From corporatism to democracy. Penguin Books, London
Sawhney M, Kaplan S (1999) Let's get vertical. Business 2.0, September 1
Schrange M (1990) Shared minds: the new technologies of collaboration. Random House, New York
Seely BJ, Duguid P (1996) 2Space for the chattering classes2. The Times Higher Education Supplement, 10 May
Simon H (1957) A behavioral model of rational choice. In: Models of man, social and rational: mathematical essays on rational human behavior in a social setting. Wiley, New York
Singh S (1997) Fermat's Last Theorem. Fourth Estate, London
Smallbone D, Welter F (2001) The disincentives of entrepreneurship in transition economies. Small Bus Econ 16(4):249–261
Smilor R, Smilor RW, Kawasaki G (2001) Daring visionaries: how entrepreneurs build companies, inspire allegiance, and create wealth. Adams Media Corporation, Avon, MA

Smith VL (2000) Bargaining and market behaviour. Essays in experimental economics. Cambridge University Press, Cambridge, MA

South Centre (2003) Information societies: towards a south perspective – Document prepared by a science expert panel at the request of the chairman of the South Centre Board to be presented at the WSIS, December

Stacey RD (1996) Strategic management and organisational dynamic, 2nd edn. Pitman, London

Stam E, Suddle K, Hessels SJA, van Stel AJ (2007) High growth entrepreneurs, public policies and economic growth. Jena economic research paper, June, No. 2007-019. Available at SSRN: http://ssrn.com/abstract=1019429

Stanford X (2002) Social capital and innovation analysis. Case study of the E100 network: foundations and assessment overview. KnowMap 2(2). http://www.knowmap.com

Stanford X (2005) The knowledge management strategy Vee: A framework for creating a new strategic direction. In: Amidon D, Formica P, Mercier-Laurent E (eds) Knowledge economics: emerging principles, practices and policies, vol. 2. Tartu University Press, Estonia

Stewart A (2001) The conversing company: its culture, power and potential. Paper presented at the 1st world conference for systematic management, Vienna, May

Storper M (1995) The resurgence of regional economics, ten years later: the region as a nexus of untraded interdependencies. Eur Urban Reg Stud 2(3):191–221

Strandburg KJ (2005) Curiosity-driven research and university technology transfer. Paper presented to the colloquium on entrepreneurship education and technology transfer, 13 January

Sveiby KE (1997) Knowledge management and EU. challenging the perspective. In: Seminar on knowledge management and the European union – results of the workshops, Utrecht, 12–14 May

Sveiby KE (2000) Knowledge management – the viking way. http://www.sveiby.com/articles/vikings.htm

Tapscott D (1995) The digital economy: promise and peril in the age of networked intelligence. McGraw-Hill, New York

Tapscott D, Lowy A, Ticoll D (1998) Blueprint to the digital economy. Creating wealth in the era of e-business. Mc-Graw-Hill, New York

Taylor J, Wacker W.m (1997) Speak the future. Wired, June

Thompson J (1999) The world of the entrepreneur – a new perspective. J Workplace Learn Employee Couns Today 11(6):209–224

Thurow L (1992) Head to head: the coming economic battle among Japan, Europe, and America. William Morrow, New York

Tödtling F, Trippl M (2004) One size fits all? Towards a differentiated policy approach with respect to regional innovation systems. Paper prepared for the conference 'Regionalisation of Innovation Policy – Options and Experiences', German Institute of Economic Research (DIW Berlin), June 4–5. Department of City and Regional Development, Vienna University of Economics and Business Administration

Tremblay K (2002) Student mobility between and towards OECD countries in 2001. A comparative analysis. OECD Directorate for Education, Paris. http://www.ulb.ac.be/unica/Tremblay_Paper.doc

Tselentis G et al (2009) Towards the future Internet. IOS Press, Amsterdam

Uglow J (2002) The lunar men: five friends whose curiosity changed the world. Faber and Faber, London

UNDP (2003), Annual Report. A World of Development Experience

UNCTAD (2005), World Investment Report. Transnational Cooperation and the Internationalization of R&D

US National Commission on Entrepreneurship (2001a) Building entrepreneurial networks, Washington, December

US National Commission on Entrepreneurship (2001b) Five myths about entrepreneurs: understanding how businesses start and grow, Washington, March

US National Commission on Entrepreneurship (2002) Entrepreneurial Culture: Talking about Culture and Entrepreneurship, No. 52, Tuesday June 11 http://www.redherring.com/insider/2002/04-09/2509.html

US National Commission on Entrepreneurship (2002a) Entrepreneurial culture: talking about culture and entrepreneurship, No. 52, June 11, Tuesday
US National Commission on Entrepreneurship (2002b) Angels to the rescue? Biweekly Report, No 62, October
US National Commission on Entrepreneurship (2002c) The return of regional development organisations. Biweekly Report, No 56, July 1
US National Commission on Entrepreneurship (2002d) America formula for growth. Federal policy and the entrepreneurial economy, 1958–1998, Washington, October
Uzzi B (1997) Social structure and competition in interfirm networks: the paradox of embeddedness. Adm Sci Q 42(1):35–67
Vallat J (2009) Intellectual property and legal issues in open innovation in services. OISPG, Brussels
van der Speck R, Spijkervet A (1997) Knowledge management. Dealing intelligently with knowledge. Kenniscentrum CIBIT, Utrecht
van Stel A, Carree M, Thurik R (2005) The effect of entrepreneurial activity on national economic growth. Discussion papers on entrepreneurship, growth and public policy. Group Entrepreneurship, Growth and Public Policy, Max Planck Institute for Research into Economic Systems, Jena, Germany. http://d.repec.org/n?u=RePEc:esi:egpdis:2005-04&r=ent
Verespej MA (1999) Invest in people. They are the only sustainable competitive advantage. http://www.industryweek.com, March 2
Voice of the Entovation 100 (2002) Focus on new products/services, mimeo. http://www.entovation.com
Watson W (2002) Europe uber alles? Don't bet on it. National Post, September 24
Wells J, Barley K (1998), Corporate Universities: Partners in Workforce Education. Paper delivered at the "European Workforce Development Conference", Nottingham: University of Nottingham, 11–13 May
Wiig KM (2005), Stories, mental models and situation-handling. In: Amidon D, Formica P, Mercier-Laurent E (eds) Knowledge economics: emerging principles, practices and policies, vol. 1. Tartu University Press, Estonia
Wills J (2001) 1688: A global history. Granta Books, London
World Economic Forum (2002), The Lisbon Review 2002-2003. Assessment of Policies and Reforms in Europe, Geneva, www.weforum.org
World Bank (2008) World development indicators. World Bank, Washington, DC
Wyllie J (2002) Thomas Edison's business model, mimeo.
Yamazaki H (2004) Knowledge communities in Japan: a case study. Knowledge Board, March 10. http://www.knowledgeboard.com
Yin D, Lin J (2002) Sharing tacit knowledge in Asia. Knowl Manage Rev 5(3), July–August
Zeldin T (1994) An intimate history of humanity. Sinclair-Stevenson, London
Zhou G (2005) Innovated to develop: the practice of China's science parks. In: XX IASP world conference on science and technology parks in conference proceedings, Beijing

About the Authors

Thomas Andersson President of Jönköping University over the past five years and currently full Professor of International Economics and Industrial Organisation at Jönköping International Business School (JIBS), holds a number of international board and advisory positions. He is Chairman of the International Entrepreneurship Academy (Intentac), Chairman of the International Organisation for Knowledge Economy and Enterprise Development (IKED), Senior Advisor of Science, Technology and Innovation Policy at the Research Council, the Sultanate of Oman, and Vice Chairman of Division XI of the Royal Swedish Academy of Engineering Sciences (IVA) on Education and Research Policy. He is currently a member of two expert groups to the European Commission, on "The Role of Community Research Policy in the Knowledge-based Economy" and "World Class Research Infrastructures" respectively, member of the Advisory Board of the Swedish Agency for Higher Education, board member of the Swedish Programme on ICT in Developing Regions (SPIDER), and serves on the Steering Committee of the Global Forum. He was Vice President of the Italian-based International Network for Small and Medium-Sized Enterprises (INSME) from 2003 to 2007, and member of the International Advisory Board of the World Knowledge Forum, Korea, from 2000 to 2005.

Thomas Andersson has previously been Deputy Director of Science Technology and Industry at the Organisation for Economic Cooperation and Development (OECD), where he headed the technology part of the OECD Jobs Study, co-coordinated the OECD Growth Study, and run a joint program with the World Bank on Building Knowledge-Based Economies, among other activities. He has also been Assistant Under-Secretary and head of the Structural Policy Secretariat in the Ministry of Industry and Commerce in Sweden. Prior to that, he headed the international research programme at the Industrial Institute for Economic and Social Research (IUI) in Stockholm. Having his PhD from the Stockholm School of Economics where he also became Associate Professor, he has published widely and been a visiting fellow at Harvard University, Bank of Japan, Hitotsubashi University, and the University of Sao Paulo. He was awarded the annual lifetime "Distinguished Scholar Award" at Innovations'07 in Dubai.

Martin Curley is Senior Principal Engineer and Global Director of IT Innovation at Intel Corporation managing a network of IT Innovation centres catalyzing worldwide IT Innovation. He is also Director of Intel Labs Europe whose mission is to advance Intel research and innovation in Europe while partnering with the broader European research and business ecosystem to enable European competitiveness. Previously Martin has held a number of senior IT Management positions for Intel and held management and research positions at General Electric and Philips. Martin has a degree in Electronic Engineering, a Masters in Business Studies from University College Dublin, Ireland, and a PhD in Information Systems from the National University of Ireland, Maynooth. Martin is author of "Managing Information Technology for Business Value" published by Intel Press, January 04, co-author of "Managing IT Innovation for Business Value" published in 2007 by Intel Press and has published widely on a variety of technology related topics. Martin is also Professor of Technology and Business Innovation at the National University of Ireland, Maynooth and co-Director of the Innovation Value Institute, helping lead a unique industry-academic open innovation consortium to advance IT management and innovation. He is a frequent international keynote speaker on Innovation and Technology and has twice been a visiting scholar at MIT Sloan. Martin is a fellow of the Institution of Engineers of Ireland and the British Computer Society.

Dr. Piero Formica is Dean of the International Entrepreneurship Academy (www.intentac.org) and Professor of Economics with special focus on innovation and entrepreneurship at the Jonkoping International Business School. He is also Special International Professor of Knowledge Economics and Entrepreneurship, School of Economics and Management – Beijing University of Aeronautics and Astronautics, Scientific Director of the Higher Education Programmes at COFIMP (the Higher Education Institution of the SMEs in Bologna, Italy), and Visiting Professor of Knowledge Economics and Entrepreneurship at the Jean Monnet Faculty of Political Studies (Second University of Naples). Between 2003 and 2006 he held the Marie Curie Professorship at the Faculty of Economics and Business Administration, University of Tartu, Estonia. Between 1982 and 2003 he held the professorship of Economics of Innovation in the Masters of Business Law and Technology Management programmes at the University of Bologna, Italy.

Professor Piero Formica has over 30 years of experience in the fields of international economics and economics of entrepreneurship and innovation, working with OECD Economic Prospects Division in Paris, academic institutions, large corporations and small companies, governmental bodies, and the European Union.

His advice has been sought by diverse organisations such as Xerox Corporation - Xerox Professional Document Services, INTEL-Innovation Value Institute, TELECOM Italia Spa, the British Council, the European Commission (DGXIII, Science Parks, RITTS-Regional Innovation and Technology Transfer Infrastructures and Innovation Programme; DG XVI, RIS-Regional Innovation Strategies; DG XII, Evaluation), The European Business & Innovation Centre Network (Brussels), the DATAR (the French Government *DELEGATION A L'AMENAGEMENT DU TERRITOIRE ET A L'ACTION REGIONALE*), the Institute for Enterprise and

Innovation (University of Nottingham), the Institute of Competitiveness, and the public authorities in Australia, India, China, and the Middle East.

His presentations in the field of entrepreneurship and innovation have been heard throughout Europe, Asia, Australia, Latin America, Canada, North Africa, and the Middle East.

He serves as board member of Industry & Higher Education and Editorial Director for the Knowledge Economy Series published by EffeElle Editore, Italy.

In the last five years Piero Formica has extensively published in the fields of knowledge economics, entrepreneurship and innovation. Among the most recent titles:

KNOWLEDGE MATTERS: Technology, Innovation and Entrepreneurship in Innovation Networks and Knowledge Clusters (with Elias Carayannis), MacMillan Palgrave, 2008

"Innovation Networks and Knowledge Clusters in the Glocal Knowledge Economy and Society: Insights and Implications for Theory and Practice", (edited with Elias Carayannis), *International Journal of Technology Management*, Vol 46, Nos. 3/4, 2009

Le vie dell'innovazione (The Road to Innovation), Editrice Compositori, Bologna, 2009

Index

FSC
www.fsc.org
MIX
Papier | Fördert
gute Waldnutzung
FSC® C083411

Kolibri 360 GmbH
Ferdinand-Jühlke-Straße 7
99095 Erfurt, Deutschland
produktsicherheit@kolibri360.de